Coley's Toxins: The Sum of Our Hope

"All therapeutic cures are obtainable only by the working of physiological forces and the first hope of therapeutic success comes with the observation of the efficiency of unaided nature to accomplish cure...These cases, rare though they be, are the sum of our hope."— A. Pearce Gould, MD, 1910[1]

Disclaimer

This is a report on the use of complementary and alternative medicine (CAM) for the treatment of cancer. It is written for people seeking non-technical information on treatments that are not generally offered by conventional oncologists in the English-speaking world. The author is a science writer, not a medical doctor, and therefore his only aim is to describe—**not to advocate, prescribe or recommend**—the treatments in question. He and his colleagues at Cancer Communications, Inc. (CCI), urge all cancer patients to be under the care of qualified and licensed physicians, such as board-certified medical oncologists or other appropriate specialists. In particular, inclusion of a treatment, clinic or practitioner in this report does not constitute an endorsement, nor does the author or CCI share any responsibility should the reader decide to utilize one of the methods or practitioners mentioned herein. Nor is there any warranty, stated or implied, as to the accuracy or currency of the information contained herein.

Acknowledgments: I wish to thank the following for reading and commenting on this paper: David Hess, PhD, Troy, NY; Uwe Hobohm, MD, Giessen, Germany; Stephen Hoption Cann, PhD, Vancouver, BC, Canada; Ralf Kleef, MD, Vienna, Austria; Donald H. MacAdam, Ancaster, Ontario, Canada; Marge McAvoy and Liz Shapiro of the Sharon Historical Society, Sharon, CT; Jacques Miller, Melbourne, Australia; and Arno Thaller, MD, Markt Berolzheim, Germany. All remaining errors and omissions are my sole responsibility.

Table of Contents

Disclaimer

This is a report on the use of complementary and alternative medicine (CAM) for the treatment of cancer. It is written for people seeking non-technical information on treatments that are not generally offered by conventional oncologists in the English-speaking world. The author is a science writer, not a medical doctor, and therefore his only aim is to describe—**not to advocate, prescribe or recommend**—the treatments in question. He and his colleagues at Cancer Communications, Inc. (CCI), urge all cancer patients to be under the care of qualified and licensed physicians, such as board-certified medical oncologists or other appropriate specialists. In particular, inclusion of a treatment, clinic or practitioner in this report does not constitute an endorsement, nor does the author or CCI share any responsibility should the reader decide to utilize one of the methods or practitioners mentioned herein. Nor is there any warranty, stated or implied, as to the accuracy or currency of the information contained herein.

Acknowledgments: I wish to thank the following for reading and commenting on this paper: David Hess, PhD, Troy, NY; Uwe Hobohm, MD, Giessen, Germany; Stephen Hoption Cann, PhD, Vancouver, BC, Canada; Ralf Kleef, MD, Vienna, Austria; Donald H. MacAdam, Ancaster, Ontario, Canada; Marge McAvoy and Liz Shapiro of the Sharon Historical Society, Sharon, CT; Jacques Miller, Melbourne, Australia; and Arno Thaller, MD, Markt Berolzheim, Germany. All remaining errors and omissions are my sole responsibility.

Table of Contents

Table of Figures

The Story of Frank Fowles

In 1955, Frank Fowles of Ogden, Utah, was on top of the world. He owned a thriving insurance agency and represented his district in the state senate, as he had for 20 years. He was being mentioned as a potential candidate for Governor.[2] But that November, at the age of 63, he began to experience chest pains, shortness of breath, fever and chills. His doctors diagnosed him as having gallstones but upon operating, found instead pancreatic cancer, which had spread to his liver and bile ducts. They concluded that he had only 90 days to live.

Under the conventions of the time,[3] his wife and children were given the frightful diagnosis, but Frank himself was left in the dark. Instead, he was told that his surgeons had successfully removed his gallstones and that the drainage tube sticking out of his side was to remove some excess fluids. Frank returned to his daily routine, feeling generally healthy but puzzled by the behavior of those around him. "I thought there was something wrong with my wife," he later said. "She seemed to be going to pieces, and I had no idea why."

As time went by, however, things did not go as expected. He did not show any signs of the wasting syndrome that is commonly seen in advanced pancreatic cancer patients. Instead, he continued to feel well and began to nag his doctors about removing his drainage tube. Eighteen months after his surgery, the same doctors went back in to remove the tube. It was then that they found something that gave them a shock…or, rather, *failed* to find something, namely Frank's cancer. They search in every anatomical hiding place. "There wasn't a trace," they said. "We looked everywhere." Fifteen months after that, when Frank's story was featured in *Time* magazine, there was still no evidence of cancer. It had definitely been there, but he was now cancer-free. No medical intervention had "cured" him, but he was definitely cured. How this had happened is still a matter of intense curiosity and debate.

Spontaneous Remissions

Doctors call such events "spontaneous remissions." These unplanned recoveries are rare, and when they do occur, become the stuff of legend. Stories like this give hope to cancer patients. They have been called the experiments (or, more poetically, the whispers) of nature.[4,5] If accompanied by religious ritual, they may be ascribed to divine intervention, such as alleged occur at the famous Catholic shrine in Lourdes, France.[i] Others feel that

[i] "A poor woman suffered from a terrible cancer on the mouth; she was taken to Lourdes, and my friend saw her plunged into the bath still afflicted with this dreadful disease. When she came out it was gone, and only a red mark was visible to show where it had been." Anonymous. Catholic notes. *The Ave Maria.* 1883;19:276. The shrine maintains a medical department to document such therapeutic effects.

these incidents provide scientific clues about how we might treat cancer, since apparently the body sometimes produces a cure through its own mysterious devices.

Spontaneous remissions of intractable diseases—so-called "cures without cause"[6]—have probably always existed, but began to appear, with increasing frequency, in the medical literature of the 18th century.[7] Tumors sometimes diminished in size, or even disappeared entirely, without any curative medical intervention.[8] Although there was no single precipitating cause, sometimes these remissions followed an acute infection, especially one that produced high fevers and thick, opaque yellowish secretions. Doctors applied an old term to these secretions, laudable pus.[ii] It was considered laudable (or praiseworthy) because its appearance was thought to portend a good prognosis in almost any disease.[9]

A Cautionary Note

In speaking of old cases of spontaneous remission, a word of caution is in order. In assessing any particular case, we must take into account the state of histopathology (i.e., the science of detecting disease through the microscopic examination of tissue slides) that existed at the time of their occurrence. Until the mid-19th century the science was virtually nonexistent and cancer was diagnosed by its symptoms and the gross examination of tumors. In the late 19th century, microscopic studies had become much more reliable. In the 21st centuries errors in diagnosis are rare, and difficult cases are usually submitted to multiple experts for second or even third opinions.[10] Most scientists need a firmly established histological diagnosis before they will accept the idea that a person has cancer.

Rohdenburg Study

The scientific study of spontaneous remission was slow in developing. There was a rare journal survey on the topic by George Louis Rohdenburg, MD (1883-1967), in 1918.[11] Rohdenburg, a pathologist at the Crocker Research Laboratory of Columbia University, New York, examined 302 such cases, mostly cancers of the uterus and breast. He noted that remissions sometimes occurred after high fevers, confirming the link that had been noted in medical reports for hundreds of years.

In some quarters, however, talk of "spontaneous regression" or remission was seen as problematical, since it allegedly encouraged non-scientific approaches to serious medical problems. Some scientists called such stories old wives tales. Cancer was defined as a

---

[ii] Laudable pus was considered a favorable prognostic factor as early as the *Practica Chiurgiae* of Roger of Palermo (a.k.a. Ruggiero Frugardi), around 1180 CE. See: Garrison FH, *An Introduction to the History of Medicine.* Philadelphia and London: W.B. Saunders, 1914, p. 102. Surgeons would generally encourage infections in wounds and a so-called "irritative fever." This approach lasted 700 years and only died out with the adoption of aseptic surgery in the late 19th century. See: Davis L, Sabiston DC, Sabiston DC, Jr., et al. *Textbook of Surgery: The Biological Basis of Modern Surgical Practice.* Philadelphia: Saunders, 1972.

"progressive" disease, because of the belief that it was invariably fatal, unless of course surgeons intervened with heroic operations. To even mention the possibility of a spontaneous cure was, to quote one doctor, "scientifically untenable." He concluded: "Current belief is that all cancers are irreversible, that all observations of spontaneous regressions are to be dismissed as mistaken diagnoses."[12]

This negative attitude was so widespread, according to psychiatrist Gotthard Booth, MD (1899-1975), that when such cases occurred they were rejected by the editors of peer-reviewed journals, "because physicians and journals are notoriously reluctant to become associated with 'far out views."[13] Then, in a classic vicious cycle, the lack of scientific documentation reinforced the impression that there was no scientific basis to these reports.

If a patient were cured by means other than the conventional (or no means at all) this meant, to some people, that the original diagnosis must have been faulty. Doctors were sometimes known to change their first diagnosis of cancer when confronted with an inexplicable positive outcome.[iii] In those days, the "cure" of cancer by non-surgical means could easily be called a misdiagnosis, in which the patient never had cancer in the first place. This became awkward when outstanding pathologists stuck by their original diagnoses and their skill was therefore called into question.

Starting in the 1940s, another type of explanation began to emerge. Confronted by some rigorously documented cases that had benefited by unconventional treatment, doctors were forced to admit that the patients in question had been cured. Since that, a priori, could not be the result of an unconventional remedy, it must therefore have been the result of a "spontaneous remission." The problem here was the general medical prejudice against the existence of spontaneous remissions. Such claims, when they were made, met with "widespread medical skepticism," according to Dr. Booth.[14] So either spontaneous remissions did exist or scientists had to confirm the possibility that the alternative treatments were working. Spontaneous remission seemed like the more acceptable of the two alternatives.

Cole and Everson

Warren H. Cole, MD (1898-1990), was a celebrated surgeon of the mid-20th century. Together with a young University of Illinois colleague, Tilden C. Everson (1919-1994),

---

[iii] For example, in December 1907, Charles C. Scott, MD, published a case report in *General Practitioner* on the successful treatment of cancer of the caecum with digestive enzymes, whereon the operating surgeon, Prof. Albert Carless (1863-1936) of King's College, London, appended a note saying that he now doubted his own diagnosis. His reasoning was that since the treatment in question was worthless, the diagnosis itself must have been faulty, since the patient recovered. See: Moss, RW. The life and times of John Beard, DSc (1858-1924). *Integr Cancer Ther.* 2008;7:229-251.

he set about to prove that spontaneous remissions actually occurred and therefore might explain the otherwise inexplicable cures attributed to unconventional treatments, particularly one called Krebiozen.[iv] Scouring the literature, Everson and Cole eventually documented 125 cases in which "far advanced cancers seemed to cure themselves, at least temporarily." They published a 560-page monograph on the topic.[15] Cole eventually drew the lesson from this study that the "body's defense forces are mobilized against" cancer. "What we need to learn therefore is how to mobilize this force when cancer strikes."[16]

Figure 1 Warren H. Cole

Source: American Cancer Society

Everson and Cole's monograph was followed by a handful of other studies. In 1974, the Johns Hopkins Medical Institution, Baltimore, held a conference on the subject, which was then summarized in William Boyd's book *Spontaneous Regression of Cancer*. In addition, Brendan O'Regan and Caryle Hirschberg, worked on a spontaneous remission project for the Institute of Noetic Sciences. They identified and published more than 1,000 cases in the world literature.[17] This comprehensive bibliography became the basis of a popular book, *Remarkable Recovery* by Ms. Hirschberg and Marc Ian Barasch.[18]

The president of Memorial Sloan-Kettering Cancer Center in New York City, Lewis Thomas, MD (1913-1993), summed up late 20th century opinion: "Rare but spectacular remission of cancer persists in the annals of medicine.... No one doubts the validity of the observation..."[19] There is irony in the phrase "no one doubts," since for about 100 years almost all doctors had indeed doubted, with devastating effect.

[iv] Cole, who went on to become president of the American Cancer Society (1960), was the head of the "research validation committee" at the University of Illinois that investigated the disputed cancer treatment, Krebiozen.

Spontaneous Remission of Breast Cancer

More recently, H. Gilbert Welch, MD, of Dartmouth Medical School, added a new dimension to the debate. In a carefully reasoned 2008 article in the *Archives of Internal Medicine*, Welch argued that spontaneous regressions might be considerably more common than previously thought. The study followed a group of almost 110,000 Norwegian women who had undergone periodic mammographic screening for breast cancer over a 5-year period between 1992 and 1997. It compared these women with a second, matched group of women who did not undergo regular routine breast cancer screening during this same period. Surprisingly, the women who underwent regular screening had *22 percent more invasive breast cancers* than those who did not.[20]

Welch suggested a plausible explanation for this discrepancy. Women in the unscreened group had had an equivalent number of cancers, but in the absence of screening those cancers went *undetected,* and therefore *untreated.* Then, rather than progressing, those cancers simply regressed and disappeared over time. In other words, although it seems counterintuitive, up to one fifth of all breast cancers that were detected by mammography resolved themselves without medical treatment!

As the authors put it in more technical terms: "Because the cumulative incidence among controls never reached that of the screened group, it appears that some breast cancers detected by repeated mammographic screening would not persist to be detectable by a single mammogram at the end of 6 years. This raises the possibility that the natural course of some screen-detected invasive breast cancers is to spontaneously regress."[21]

Or more simply: "We believe that there are 'pseudo' cancers in this population [of screened women, ed.] and we suggest that most of these pseudo cancers regress," said Jan Maehlen, MD, professor of pathology at the Ulleval University Hospital in Oslo, Norway, and one of the study's authors, in an online interview.[22]

Most oncologists see few, if any, cases like that of Frank Fowles in a lifetime of practice. When such a case occurs, it may become the basis of a medical journal article. Sometimes, there is no apparent reason for the cure. As far as we know, Frank Fowles' tumor disappeared without ever revealing the cause of its departure. But if you look at the totality of the cases, a pattern does emerge. The most common predisposing circumstance (in roughly half the cases) is a concurrent infection with a high fever, often with the microbe *Streptococcus pyogenes*.

The Power of Fever

Ancient physicians were aware of the power of fever to cure otherwise incurable conditions. Hippocrates (460-370 BCE), the so-called father of Western medicine, spoke in praise of fever on a number of occasions.[1] The Roman physician Celsus (ca. 25 BCE-50 CE) wrote: "Fever itself is in the end often a protection, which may appear very strange."[23] In Varro's day (116-27 BCE) there were three Roman temples devoted to the goddess Febris (i.e., Fever), although their purpose is no longer known.[24]

Fever is often accompanied by inflammation. Early doctors also praised the curative power of inflammation.[25] Inflammation was viewed as "a reaction of the organism for the restoration and assertion of its integrity with increased expenditure of force of all systems."[26] The famous English surgeon, John Hunter (1728-1793), spoke of "salutary inflammation," as did many others.[27] Such observations set the stage for the findings we are about to discuss.

The Erysipelas Connection

Erysipelas is—happily—little known today since antibiotics generally cure the condition. But until the 19th century knowledge of the cause, much less the cure, was unknown. The word erysipelas was originally used by Hippocrates, who described the problem in minute and generally accurate detail in the third section of his work *Of the Epidemics.*[28] He said that erysipelas often originated from small wounds and preferentially attacked the head, which is generally true.

"Rose ersysipelas," *Kopfrose* (in German) or "St. Anthony's Fire," as it was sometimes called, was a disease of legendary virulence, a sometimes-fatal infection that could affect a particular hospital bed or rage like a fire through hospital wards.[29] When a hospital became infected with this disease, its directors were known to tear the building down rather to engage in futile attempts to eradicate the persistent organism![30,v]

But there was another side to this terrible affliction. Attentive doctors observed that erysipelas was sometimes accompanied by the rapid resolution of other diseases. "A peculiar effect is often exercised by erysipelas," wrote one professor from Berlin, "upon pre-existing chronic or acute morbid processes of the skin."[31,32]

In 1831, J. C. Sabatier of Orleans wrote in his doctoral dissertation for the Faculty of Medicine of Paris that no remedy could make syphilis disappear as quickly as a concurrent erysipelas infection.[33,34] In 1874, Charles Mauriac[vi] (1832-1905) observed an epidemic of erysipelas at the Charité hospital Paris, during which patients with syphilitic infections were put into remission of their disease.[35]

As a result of many such observations, throughout the 19th century, there were "heroic" attempts to utilize erysipelas as a treatment for non-malignant conditions, especially

---

[v] In the Mütter Museum of the College of Physicians of Philadelphia there is a preserved specimen of a case of erysipelas that still awes visitors.

[vi] Fehleisen refers to an otherwise unknown "Dr. Maurice." But Charles Mauriac was one of the top French experts on syphilis of the day, who wrote a dozen books on the topic. He is almost certainly the author in question.

syphilis. Prof. Fehleisen tells us[36] that Philippe Ricord (1800-1889)[vii] and Armand Després (1834-1896) made the "first attempt to use erysipelas as a curative agent," in some cases of syphilis.[37]

Figure 2 Philippe Ricord

Source: National Library of Medicine

Wilhelm Busch and the Cancer Connection

In 1866, Wilhelm Busch, MD (1826-1881), professor of surgery at the University of Bonn, was treating a 43-year old woman who had numerous firm sarcomata on her face and head. Some of these malignant growths were as large as pigeon's eggs. After attempting to remove a portion of these tumors, she developed erysipelas (not an uncommon result of operations in those pre-Listerian days.) But, to everyone's amazement, under the influence of this skin infection, all of her many tumors disappeared, somehow "reabsorbed" by her body.[38]

[vii] This famous French physician was actually born and raised in Baltimore, but returned to his parents' native country to study and practice medicine.

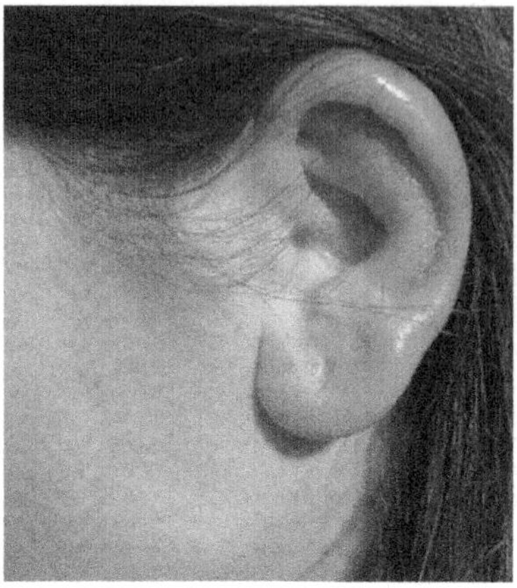

Figure 3 Erysipelas of the ear

Source: Wikimedia Commons

As we have shown, spontaneous cures of cancer had been observed before. There was also literature on the opposition of erysipelas to diseases other than cancer. But Busch's interest was prodded by a second such remission under the influence of erysipelas.[39]

Inspired by these erysipelas-induced remissions, Busch had the brilliant but risky idea of harnessing erysipelas as a cancer treatment. It is possible he knew of the earlier attempts by Ricord and Després to do a similar thing in cases of syphilis. Nevertheless, it required great daring and imagination on his part to see past erysipelas's destructive threat to its therapeutic potential in cancer. What Busch needed, however, was a likely patient with very advanced disease who was willing to undertake such a risky procedure.

His opportunity came two years later, in the summer of 1868, when a suitable patient entered his clinic, a 19-year-old with an enormous sarcoma in the lymph nodes of her neck ("mit einem gewaltigen Sarcome der Halsdrüsen," wrote Busch).

Even today, such sarcomas are difficult to treat, and in those days, except for radical surgery in the most favorable cases, there was little that could be done. Over a period of five months this tumor grew with legendary rapidity, until it was the size of a child's head. The woman's breathing was now threatened and she could no longer open one eye.[40] Busch knew that working with erysipelas posed a danger to all concerned (himself included). But he was encouraged by the two spontaneous remissions of previous years.

According to his classic paper on the topic, he first burned the skin over the tumor and attached a bandage from his erysipelas patient onto that burn. Nothing happened. There was a corner of the clinic with a bed, where previously many patients who had open wounds had become infected with erysipelas. Normally, only patients who did not have open wounds were placed there. Busch now deliberately put the young cancer patient into that so-called erysipelas bed. He again burned a coin-sized lesion over the tumor using a hot iron. After one week, she developed a fever of 40° C (104° F) and typical signs of erysipelas. The tumor, which had previously been firm and elastic, became softer and doughy in consistency.[41,42] One tumor close to her spine disappeared completely, while the main tumor shrank to the size of an apple within two weeks. Unfortunately, the woman then developed circulatory problems. Busch interpreted that to mean that her body was having difficulty resorbing copious dead tumor tissue (what we today might classify as tumor lysis syndrome). He therefore reluctantly stopped the treatment and we are uncertain what next happened to the patient.

This regression, temporary though it probably was, was a landmark in the history of cancer immunology. We must remind ourselves that a tumor regression of any sort before the modern era was considered a miraculous event.[43, 44] Outside of radical surgery (which was often ineffective and even fatal in its outcome) there was little that could be done for most cancer patients. Most non-surgical treatments were more in the realm of magic than of science. Prof. Busch's erysipelas experiment, primitive as it now seems, was in retrospect the start of a more scientific way of approaching cancer therapy.

Figure 4 Wilhelm Busch (1826-1881)

Source: the Art Collection of the Charité Hospital, Humboldt University of Berlin[viii]

One can imagine what his fellow doctors thought as this professor deliberately infected an afflicted cancer patient with another deadly disease.[45,46,47] Needless to say, such a drastic experiment would be unlikely to pass muster with institutional review boards and

[viii] http://www.sammlungen.hu-berlin.de/dokumente/8365/

ethics committees today! But, medical doctors, especially university professors, had no such constraints in those days.

Who was Wilhelm Busch, who made such an epochal (albeit little noted) discovery in the history of medicine? Information is scant. He was born in Marburg, Germany, on January 5, 1826. In 1848 (a year of revolution across Europe) he received his medical degree from the Friedrich-Wilhelms University in Berlin. There he studied comparative anatomy with the naturalist Johannes Muller and his doctoral dissertation was *On the Brains of Sharks, Rays and Sturgeons* (*Über das Gehirn der Haifische, Rochen und Störe*).[ix] Busch moved quickly up the academic ladder. In 1852 he received his Habilitation (a postdoctoral degree) in surgery and from 1852 to 1855 was a Privatdozent (a kind of associate professorship) in surgery at Friedrich-Wilhelms University. During this period (1851-1853) he served as assistant to the military surgeon, Bernhard von Langenbeck, MD. In 1855 Busch assumed the post of Professor of Surgery at the University of Bonn, where he served until his death at age 55 on November 24, 1881.

Busch's contribution was great, but was later overshadowed by American developments in this field. Clearly, however, he was a man of great daring whose contribution to modern medicine, however profound, remains underappreciated.

Fehleisen and Others

Busch may have been the first but was not the only German doctor to attempt to utilize erysipelas against cancer. Prof. Friedrich Fehleisen (1854-1924) of the University of Würzburg showed that erysipelas was caused by a bacterium in 1883. This microbe was eventually found to be identical to *Streptococcus pyogenes.*[48] (*Pyogenes* means pus producing, from the Greek word for pus, *puon).*[49,50] *Streptococcus pyogenes* was an anaerobe, growing in the absence of oxygen, and was part of the normal flora of the human nose and mouth. Acute 'strep' infections generally took the form of sore throats (pharyngitis), scarlet fever and, as stated, erysipelas. Under certain circumstances, it could also cause pneumonia, scarlet fever and rheumatic fever.

Fehleisen had a similar therapeutic idea as Busch. But, thanks to him, medicine now had at its disposal pure cultures of erysipelas's causative organism, not just "erysipelas beds" or swabs of laudable pus. Fehleisen was the first to use *cultured bacteria* in an attempt to combat cancer. Because of his international reputation as the discoverer of the bacterial cause of erysipleas, his word carried considerable weight, even among skeptics.

Fehleisen wrote:

[ix] By an odd coincidence, the biology of elasmobranch fish (sharks, skates and rays) was the scientific starting point for another great 19th century cancer pioneer, Dr. John Beard, whose research gave rise to the enzymatic treatment of cancer. See: Moss, RW. The Life and Times of John Beard D.Sc. (1858-1924). *Int Ca Ther.* 2008;7:229-251.

> "It is undeniable that many tumors have undoubtedly disappeared by the action of erysipelas. Mistake is not possible as to this, because such degeneration of new formations does not otherwise occur. Many swellings of the skin, epithelioma, keloid, carcinoma of the mamma [breast cancer, ed.], and lymphatic gland enlargements of various kinds, have been partially absorbed as a result of an attack of erysipelas."[51]

Fehleisen had some successes, as well as failures, but eventually discontinued this work.[52, 53] Around the same time Paul von Bruns (1846-1916), a surgeon in Tübingen, Germany, also observed a remission of a recurrent and widespread melanoma under the influence of erysipelas. This patient remained entirely well for at least nine and a half more years. Intrigued, Von Bruns then surveyed the use of erysipelas as a treatment for cancer and concluded that it was effective at causing regressions in a large minority of cases.[54] He surmised that these spontaneous remissions had something to do with stimulation of the immune system, which was becoming a topic of interest in scientific laboratories.

There were similar experiences elsewhere.[55] Paul Friedrich Richter, MD, who worked in the third medical clinic of Berlin, used cultured *Streptococcus pyogenes* to treat leukemia. Johannes T.W. Petruschky, MD (b. 1863) who directed the bacteriological laboratory in Danzig, treated 38 cases of inoperable breast cancer with *Streptococcus pyogenes.* He was able to produce a violent case of erysipelas in 11 of the 38 cases, within a comparatively short time. However he only could achieve stabilization, not cure. "No visible advance of the disease could be observed externally during this time, although none of the cancerous nodules disappeared and there was a decided failing of the general strength," he wrote.[56,57]

Others who examined the link between erysipelas and cancer during this period were R. Volkmann, (1830-1889)[58] Ricord, Depres, and Kleebat, all of whom "employed cultures of the erysipelas coccus for inoculating malignant tumors, with results more or less favorable."[59]

In 1897, Rudolph Eschweiler (1870-1932), a professor at the University of Bonn, published a monograph in which he described in great detail the literature on erysipelas and malignant tumors.[x] Eschweiler collected a total of 69 cases: 27 sarcomas, 38 carcinomas and 4 of indeterminate type. Of the 27 sarcomas, 9 were apparently cured (33.3 percent) while in 4 other instances the patient remained well for from two to eight years, for a total response rate of 48.1 percent.[60]

---

[x] Eschweiler is described in various sources as a professor of pathological anatomy and of ear, nose and throat medicine. Only 2 copies of this book exist in libraries worldwide: one at the Countway Library of Harvard University, the other at the University of Chicago. There is a summary of his results in Bradley L. Coley, Mixed bacterial toxins (Coley's serum). In: Pack, George T. and Livingston, Edward T., Eds. *Treatment of Cancer and Allied Diseases.* New York: Paul A. Hoeber, 1939, vol. 3, pp. 2480-2483.

The curative effect of erysipelas on cancer—what the French called "erysipele salutaire"—was sufficiently well known that the celebrated Russian author Anton Chekhov (1860-1904), himself a medical doctor, mentioned it in his letters and diaries.[61] "It was observed long ago that with the development of erysipelas, the growth of malignant tumors is temporarily checked," he wrote.[62]

Julius Wagner Jauregg

One celebrated practitioner of the time who may have influenced the development of bacterial vaccines was Julius Wagner Jauregg, MD (1857-1940). Jauregg was a celebrated Viennese neurologist. Starting in 1887, he used erysipelas to treat mental illnesses. Then, in 1917, he began inoculating live malaria parasites into patients who were suffering from dementia paralytica (also called general paresis of the insane), caused by neurosyphilis.[63] His results were thought to be quite successful and he was awarded the Nobel Prize in Medicine in 1927.

Enter William B. Coley, MD

The young doctor who was to introduce the erysipelas treatment to America, and is the main subject of this essay, was initially unaware of the abovementioned European reports. Like Busch, von Bruns and others, he initially relied on his own fortuitous observation of an erysipelas-induced remission. The name of this doctor was William B. Coley, MD (1862-1936), and he is now universally regarded as the 'grandfather of cancer immunotherapy.'[64] Over a period of 45 years, he did more to establish the benefit of bacterial products than any other individual. It is therefore proper to consider his life and achievements in some detail, for the light they shed on the historic quest for an effective biological method of treating cancer.

William Bradley Coley was born in Saugatuck, the northern section of Westport, Conn., on Jan. 12, 1862. He came from distinguished New England forebears. On his father's side, he was in the ninth generation of descent of Samuel Coley, who in 1631 had immigrated from Hertfordshire, England to the Massachusetts Bay Colony. Samuel was among the earliest settlers of Milford, Connecticut (1639) and his son Peter Coley was an early settler of Fairfield County. [65]

On his mother's side, William was descended from Samuel Wakeman, who came to America in the same year as Samuel Coley and was one of the original settlers of Hartford, Connecticut. Samuel Wakeman's son, John, was an early member of the New Haven colony, attended Harvard College and became the second Congregational minister of Fairfield, Connecticut.[66] The Coleys were part of the stout yeomanry that settled the American colonies. Various ancestors fought in the Revolutionary and Civil Wars and in William B. Coley's youth, his cousin, Joseph P. Bradley (1813-1892), was Chief Justice of the United States Supreme Court. (See Appendix A for genealogical information on the Coley family.)

The Coleys were farmers with an intellectual streak. Dr. Coley's father, Horace Bradley Coley, not only worked his 65-acre parcel but was also the local schoolmaster. His brother, the Reverend James E. Coley, ran a private secondary school nearby, which William B. Coley later attended. The family lived in a settlement called locally "Coleytown," containing (according to census records) about 20 of his close relatives. Although the name Coleytown never attained official status, and the "town" itself disappeared, the name survives today as the local name for the elementary and middle schools and the fire station (see Figure 2).

Figure 2 Coleytown Elementary School, Westport

Dr. Coley spent his youth in this rural setting of gently rolling hills. But his early years were marred by the sickness and premature death of many of the beloved figures in his life. His 26-year old mother Clarina died of typhoid fever a few months after his birth. His father then married Abbie Gray, who raised young William until her death in November 1879. So, it has been said, Coley lost two mothers before he even turned 18.

In his excellent history of cancer immunotherapy, *Commotion in the Blood*, author Stephen Hall expanded on this series of personal tragedies:

> "Of Coley's first two stepsisters, one later died of peritonitis, the other of an unknown fever; Coley's sister Carrie, probably the most beloved of all in his immediate family, died in 1892 after a life-threatening pregnancy and an emergency abortion late in the first trimester (which Coley reluctantly performed himself), and Coley also lost his second-born son Malcolm, who died in 1901 at age five of an acute gastrointestinal infection, after less than a week of illness."[67]

Coley's Education

Until the age of 13, Coley attended the one-roomed precursors of the present-day Coleytown public schools, probably with his father as schoolmaster. From 13 to 15 he attended Greens Farms Academy, five miles to the south, and then the private school in Westport founded by his uncle. His final year of high school was spent at the Easton Academy, studying under Prof. John MacNiel. In September 1880 he entered Yale University, where he graduated in four years with honors in Latin and Greek. In that same year, 1884, he met and courted 18-year-old Alice L. Lancaster of Newton, Mass., whose father, Charles B. Lancaster, was a wealthy Boston shoe manufacturer. William and Alice were eventually married in the summer of 1891.

Coley once wrote that he knew he wanted to become a medical doctor even before entering Yale. But, rather than enrolling in medical school immediately after graduation, he took a position as Senior Master at the Bishop Scott Grammar School in Portland, Oregon. He taught Classics there for two years (1884-1886), but in his second year he spent his spare time studying anatomy and physiology. He did so well that, upon returning east in 1886, he was allowed to skip these subjects and enroll in the second year of the (then) three-year course at Harvard Medical School, graduating in just two years.

Figure 5 Coley as a young man

In the spring of 1888, just before graduating from Harvard, Coley won a competitive examination to attain a surgical internship at The New York Hospital. As a result, starting in October 1888 he worked in the service of two of America's best known surgeons, Robert F. Weir, MD (1838-1894) and William Tillinghast Bull, MD (1849-1909). "His internship under such masters of surgery was of the greatest value to him and furnished a stimulus that continued to act for many years," according to a 1919 history of New York medicine.[68]

Coley finished his internship in June 1890 and spent the following year's residency at the Hunt Memorial Hospital in Port Jervis, NY. On June 4, 1891, he married Alice Lancaster at her capacious home in Newton, Mass,[69] and then took up his duties at Memorial Hospital.

The couple eventually had three children, Bradley, Malcolm (who, as mentioned, died in childhood) and Helen. Bradley and Helen will figure prominently in this story. The family—compact by the standards of the day—lived at three consecutive locations in Manhattan: 5 Park Avenue, 655 Park Avenue, and 123 East 53rd St.

In April 1901 Alice Coley bought Okeden, a 15-acre estate, in rural Sharon, Connecticut. Perhaps the property's 100-year-old house reminded Coley of his own rural roots (see Figure 5). (His next door neighbor, and sometimes business partner, was a Texas oil millionaire named William F. Buckley, Sr.)

Figure 6 Coley's country home, Okeden

Photograph courtesy of the Sharon Historical Society, Sharon, CT

Coley was bright and conscientious and came from the "right" social background. Not surprisingly, he quickly moved up the medical ladder. In the spring of 1891 he was appointed instructor in surgery at New York Post-Graduate Medical School and Hospital, an appointment he held for the next six years. He then accepted the position of assistant professor of clinical surgery at the College of Physicians and Surgeons (P&S), now part of Columbia University. He continued at P&S until 1909 when he was appointed professor of clinical surgery at Cornell University Medical School, in New York City. He continued at Cornell until his retirement in 1933.

Leaving aside for a moment his work on cancer, Coley was also a recognized expert on surgery of the abdomen, particularly hernias. His base of operations for this was the Hospital for the Ruptured and Crippled (the R&C, now the world-renowned Hospital for

Special Surgery on East 70th St. in New York). He was associate surgeon (1890-1909) there, attending surgeon (1909-1924), and finally surgeon-in-chief (1924-1933).[70]

Coley also served as a surgeon for the New York Central Railroad. He was a leading member of the New York and New England Association of Railway Surgeons and became the group's president in 1920-1921.[71] He also wrote chapters on hernia for three medical textbooks and is remembered for pioneering the use of surgery to repair inguinal hernias in children.[72,73,74]

Since railroads were plagued by hernias among its track workers, he had no shortage of such patients. In 1908, he was appointed the first surgeon of the New York Central Railroad, serving as chief surgeon until 1932, when he reached retirement age. Additionally, in 1918, the Surgeon General of the Navy established a program to have hernias of military personal treated at the R&C. Coley directed this program. It was reported that to accommodate all these military personnel, the hospital had to encroach upon the children's and women's wards, the schoolrooms, the gymnasium, and in the summer, the roof! There were 120 beds provided for soldiers, sailors, and marines that were filled continuously. Coley served as a major in the Army reserves for many years.

We can gather an idea of his standing as an intestinal surgeon from a review of his 145-page contribution to *Progressive Medicine* in 1902.[75] "As Dr. Coley has had for many years a very large experience in connection with this subject," wrote the editors of *Therapeutic Gazette,* "there are few who are better qualified to speak in this respect,".[76]

At the time that Coley graduated from Harvard Medical School in 1884, a hernia operation was considered a dangerous, even life-threatening, procedure. Patients often therefore allowed their hernias to grow to enormous proportions before submitting to surgery, whereupon they often died of bleeding or infection. It was an Italian surgeon, Prof. Edoardo Bassini (1844-1924) of the University of Padua who made the first great breakthrough in tissue repair – "a radical cure of inguinal hernia," as he called it in 1887, while Coley was still at Harvard.[77]

This technique did not become known outside Italy until 1890. Soon afterwards, young Coley helped introduce these "radical" techniques to America. He showed outstanding skill. Out of 2,000 hernia operations that he performed at the R&C there were only four deaths. Of around 1,000 cases he treated in adults only one patient died—a mortality rate of one tenth of one percent. Coley was a great admirer of Lord Joseph Lister, inventor of antiseptic and aseptic surgeon. Lister, in turn, became his friend and mentor, and greatly admired Coley's work. Quoting Lord Lister,[78] William W. Keene, MD, professor emeritus of medicine at Jefferson Medical College, Philadelphia, enthused about Coley's surgical skill: "An achievement like that is enough to cause gladness in the heart of any man who loves his fellow man."[79]

Elizabeth "Bessie" Dashiell

But Coley is best rememebered for the cancer treatment that bears his name. The story of how he developed this unusual treatment tells a great deal about the man. As early as 1890, he had begun to develop a special interest in cancer. One of his first patients, in 1891, was a 19-year-old woman named Elizabeth (Bessie) Dashiell. Bessie was the childhood sweetheart of John D. Rockefeller, Jr., heir to the Standard Oil fortune.

Figure 7 Bessie Dashiell and John D. Rockefeller, Jr.

Following a bizarre accident, in which she caught and injured her arm between two seats of a railway car, she developed a painful growth that turned out, when biopsied, to be a round cell sarcoma of the bone. (It was commonly thought at the time that traumas could precipitate malignant growths.[80]) Although Coley did everything that Drs. Bull and Weir had taught him to do in such cases, including eventually amputating the young lady's arm below the elbow, this aggressive sarcoma soon metastasized to her breast and abdomen. On January 23, 1891 Bessie Dashiell died of cancer. Everyone connected to the case was shaken and changed by this tragic death. John D. Rockefeller, Jr. pledged himself to further the cause of cancer research and his family became a major source of support for Memorial Hospital to the present day.

But the effect on Coley was equally profound. Determined to find out whether there was any effective treatment that he had failed to learn about at Harvard, Coley examined the records of all the sarcoma patients who had been treated at The New York Hospital during the preceding 15 years.[81] There were about 100 such cases, about half of whose records contained follow-up histories. The grim news was that all similarly advanced patients had died. But there was one glaring exception. This was a German immigrant named Fred K. Stein, upon whom Dr. Bull had operated four times for a round cell sarcoma of the cheek. Each time, Stein's sarcoma had recurred but the fourth and incomplete removal of the tumor in October 1884 had gone awry. Instead of curing Stein Dr. Bull had left him with a case of facial erysipelas. This kind of streptococcal infection waxed and waned. Then a strange thing happened. As described in The New York Hospital records:

> "After each [attack of erysipelas], the cicatrisation [healing] of the ulcer has rapidly proceeded, and during the attacks the flabby and apparently sarcomatous granulations have been absorbed. There is now left a healthy ulcer the size of a silver dollar."[82]

In other words, to everyone's astonishment, all traces of active tumor disappeared, following this streptococcal infection acquired in the hospital. The patient was dismissed in February 1885 and then was lost to follow-up. Intrigued to know the outcome of this unusual case, in 1891, Coley took an unusual step. He determined to find Stein. This was not easy, as – like many immigrants – the man moved frequently amid the warren of immigrant housing on New York's Lower East Side. Using the little German he had learned at Yale, Coley traipsed up and down tenement stairs, inquiring after the elusive Mr. Stein.[xi] Coley had only a rough description of a long-faced, dark-haired man with a full black beard, which only partially hid the scars from his radical cancer operations.[83]

Coley eventually located Stein, who was still alive and seemingly well seven years (i.e., 1885-1892) after this unexplained regression of his tumor. Coley took him back to The New York Hospital to be examined and photographed by his original surgeon, William Bull. Examining him in every way possible, Bull affirmed that Stein was cancer-free and had remained cancer-free since those fateful days in 1885 when he had undergone his "spontaneous remission."[84]

Fascinated by this experiment of nature, Coley began to explore the scientific literature for other connections between erysipelas and cancer. It was at this point that he

---

[xi] What was Coley's inspiration for this type of medical detective work? One explanation might be that at that very moment the Anglo-American world was mesmerized by the introduction of the world's first consulting detective, Sherlock Holmes, and his medical sidekick, Dr. Watson. *The Adventures of Sherlock Holmes* was appearing in serial form in the *Strand* magazine just at the time that Coley began his quest. There is an element of Holmes-like detection in Coley's intrepid hunt for Mr. Stein. There are other links between Sherlock Holmes and medicine. Holmes's creator, Sir Arthur Conan Doyle, was himself a physician and almost certainly modeled Holmes on his own medical school professor, Joseph Bell (of Bell's palsy fame).

discovered Dr. Busch and the fairly extensive German literature on the connection. He later reviewed this in an article for the *Annals of Surgery*, giving full credit to his German predecessors.[85] "If erysipelas, a disease produced by a specific organism, could cure a case of undoubted sarcoma when occurring accidentally," Coley wrote, "it seemed fair to presume that the same benign action would be exerted in a similar case if erysipelas could be artificially produced."

Enter Mr. Zola

Coley also said he was "so strongly impressed with the [Stein, ed.] case that I determined to try inoculations in the first suitable case."[86] In April 1891, he got his chance. Dr. Bull had a patient, an Italian immigrant named Mr. Zola, who had an inoperable recurrent spindle cell sarcoma of the tonsils and glands of the neck. Dr. Bull had operated twice, but each time the cancer recurred. Coley reports that this patient was "in such a hopeless condition that he was willing to try anything." The tumor on the tonsil was in fact blocking his food intake, and so it seemed only a question of whether he would die first of metastatic cancer or of starvation.

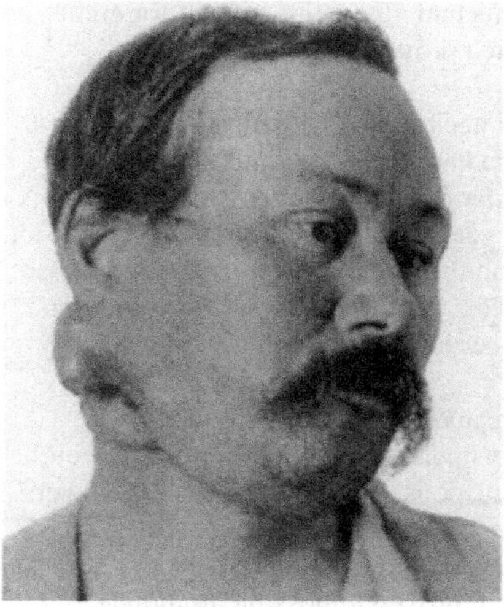

Figure 8 Mr. Zola

With the encouragement of Prof. Bull, Coley deliberately infected this patient with *Streptococcus*. Since erysipelas was such a danger, The New York Hospital refused permission to do this experiment, and so Zola was treated in his own apartment.[87,88] Zola received three applications of the toxins over a period of three weeks. There was no significant fever and so Coley then tried a higher dose. The patient then developed severe chills, headaches and vomiting, typical signs of a strep infection. However, his temperature did not reach or exceed 102° F, which is what Coley sought. But his tumor did diminish in size and, after a month or so, Zola could eat again.[89]

Dissatisfied with these palliative effects, Coley obtained a culture of erysipelas, via his friend Farquhar Ferguson, from the Berlin laboratory of "microbe hunter" Robert Koch (1843-1910).[90] This was a super virulent 'bouillon' of microorganisms. Coley injected seven minims of the fluid in the skin over the tumor on the neck.[xii] In one hour, Zola suffered a severe chill that lasted 40 minutes, followed by nausea, vomiting and severe pain. His temperature rose to 105° F (40.5° C). In 12 hours there followed a "typical attack of erysipelas," which spread over his face and neck. Coley later said that "the patient nearly died, and his niece who was nursing him also contracted the disease and nearly died."

In about ten days, this attack subsided, but the changes to the tumor were dramatic and profound. Zola's neck tumor became paler, softer and began to break down with a free discharge of what was called "caseous material" (i.e., damaged or destroyed tissue with a cheese-like appearance). This continued for another ten days. The outlying metastases disappeared without breaking down and the tonsil tumor decreased in size.[91] At the end of two weeks the tumor in the neck had "disappeared entirely," Coley later wrote.[92] The man's appetite returned and he gradually regained his lost weight. In fact, he was perfectly well. The tonsil tumor never disappeared, but remained as a hard, fibrous mass, "apparently having lost its malignant character." This patient continued well for eight more years, "when there was a local recurrence which proved fatal not long after."[93,94]

Even today, when remissions of advanced head and neck cancers are rare, such a regression would cause medical comment.[95] But one has to bear in mind that this remarkable remission occurred in 1891, i.e., *five years before* Roentgen's discovery of X rays, when no conventional treatment options were yet available. This must have been an electrifying experience for Coley, as well as Bull and the others who worked on the case. In fact, it set in motion a whole new research direction, not only for Dr. Coley but – one can say without exaggeration – for humankind, cancer immunotherapy.

Over the following two years (1891-1892) Coley treated 12 patients with inoperable carcinomas or sarcomas and succeeded in giving erysipelas to eight of them. Those eight patients who were successfully infected had tumor responses as well, two of which were complete. Thus, under the circumstances, erysipelas was proving to be remarkably effective. However, two patients died of erysipelas, which even in those days of "heroic" treatment was an unacceptable degree of risk. Also, it proved impossible to induce erysipelas in some cases.

So Coley abandoned the use of living cultures of strep. Luckily, unlike Prof. Busch, he did not abandon the treatment entirely. Instead, in January 1893 he decided to try the use of a heated bacterial extract.[96] He gently heated the serum until there were no longer any living *Streptococcus pyogenes* present. Hopefully, however, there would be enough of their byproducts left behind to have a similar effect as the living organisms. And indeed,

---

[xii] According to the *The American Medical Association's New and Nonofficial Drugs*, a standard dose of the toxins was 1 to 8 minims, or 0.05 to 0.5 cubic centimeters (cc).

even without live microorganisms, he could produce many of the same symptoms as erysipelas—fever, chills, aches and pains—without the risks (to patients, their relatives and health care workers) associated with the use of virulent microbes.

Bacillus Prodigiosus

In addition, in 1892 Coley adapted another extremely important innovation that was coming to light just at that time. He mixed the strep toxins with another bacterial byproduct, this one from killed *Bacillus prodigiosus.*[xiii]

In 1890, Georges Eugene Henri Roger (1860-1946), Professor Extraordinary at the Paris Medical School, had shown that if the streptococcus of erysipelas were grown together with this "prodigious" microbe, the virulence of the former was much increased by the latter.[97] *Prodigiosus*—or to use its modern-day scientific name *Serratia marcescens*—acted as a predisposing agent favoring the development of streptococcal infections.[98] However, Roger never himself used this mixture therapeutically.[99]

Serratia marcescens itself had anticancer activity, possibly greater than *Streptococcus* itself. This was first shown by Victor Clarence Vaughn, MD (1851-1929), dean of the University of Michigan Medical School, Ann Arbor,[100] His observations were confirmed by Coley's colleague, Martha Tracy, in 1907. Dr. Tracy and her colleague S.P. Beebe were able to make large, multiple sarcomas in dogs eventually disappear with injections of *Serratia* alone. But when they used Coley's toxins the tumors would "rapidly disappear."[101]

In the 1940s, Murray J. Shear, MD, of the US National Cancer Institute (NCI), head of chemotherapy research at the National Cancer Institute (NCI), conducted a series of experiments with *Serratia marcescens*, from which he isolated a polysaccharide he called lymphotoxin. [102] Lymphotoxin produced hemorrhagic necrosis of tumors and Shear understandably believed that he had found the "active ingredient" in Coley's fluid. However, a clinical trial of lymphotoxin showed that it caused severe side effects, too severe in this concentrated form for human use. Shear, although well intentioned, unfortunately paid little attention to the knowledge Coley imparted in his many papers on

---

[xiii] This microorganism was called "prodigious" because of its role in religious history. It sometimes infected grain products and turns them a startling red color. This was a repeated source of wonderment in the Middle Ages, as the afflicted holy wafer, believed by the faithful to be the transubstantiated body of Christ (the *Corpus Christi*), was thought to be bleeding. One persistent explanation was that local Jews had "tortured" the wafers to make them gush blood. For this reason thousands of medieval Jews were burned at the stake for allegedly "torturing" holy wafers in this way. In 1819 Italian scientists revealed that a microbe was responsible. The Italian discoverers called the microbe *Serratia marcescens*, after Serafino Serrati, Italian inventor of the steamboat. Others, however, called it *Bacillus prodigiosus*, in recognition of its religious connotations. In the end, the less evocative term was chosen and scientists now universally call it *Serratia marcescens.*

the topic, according to Prof. Hoption Cann of the University of British Columbia. "When Shear used his *Serratia* extract, he often hastened the death of his subjects because he had no concept of what symptoms should guide the doses he should use."[103] As recently as 2006, there was a phase I clinical trial with a lymphotoxin derivative in Guangzhou, China. The results however were at best equivocal.[104]

Serratia continues to be explored as a source of anticancer drugs or biological response modifiers. Not long ago it was announced that *Serratia* by itself contains "potent cytotoxic activity towards cancer cells."[105] Coley and his contemporaries knew as much 100 years ago.

The First Case

In January 1893, Coley treated his first case with the mixed bacterial vaccine that came to be known as Coley's toxins. He was J.F.F., a 16-year old sarcoma patient at the New York Post-Graduate Hospital. He had a large, inoperable spindle cell sarcoma (13 cm by 15 cm in size), extending the entire thickness of abdominal wall, attached to pelvis and involving bladder wall. His general condition was poor, he had lost control of his bladder, and he was confined to bed most of time. The disease was first diagnosed in October 1892; the boy had exploratory laparotomy and biopsy on January 24, 1893.

Coley's toxins were injected into the tumor for the next six months. The tumors steadily decreased in size, a decrease that continued even after administration of the toxins were stopped. The patient had a complete regression, with no recurrence. He continued in perfect health until his sudden death on February 28, 1919, of chronic myocarditis, 26 years after onset of sarcoma. To repeat, this extraordinary case was the first patient of any type of cancer to receive Coley's mixed toxins.

Injecting the serum in or near the tumor caused fevers as high as 40° C (104° F) or more. Previously, German scientists had observed that erysipelas's curative action seemed tied to its ability to cause fevers. The higher the fever, the greater the likelihood of cure. Observing this, Coley's thenceforth tried to deliberately and repeatedly evoke fevers in the 102 to 104° F (38.8 to 40° C) range, sometimes as high as 105° F (40.6° C), as well as chills and other flu-like symptoms in cancer patients. Eventually, as more was learned about human immunity, it was realized that the crypto-infections induced by the toxins powerfully stimulated the immune system, particularly the macrophages and lymphocytes (white blood cells). But understanding of the effects only evolved slowly over the course of a century.[106] The mechanism continues to generate debate.[107,108]

Coley's treatment not just caused tumor softening or regression, but sometimes brought about other beneficial effects as well: pain reduction, mood improvement, appetite and weight gain, more rapid wound healing, reduced edema, and sometimes even a regeneration of bone, in cases in which the bone had been destroyed by the tumor or removed during surgery. But, again, such beneficial effects were most common when fever was repeatedly provoked. In other words, even when not curative (which was a tall order in the very advanced cases that he undertook to treat) his treatment could still have valuable palliative effects. Improvement was often rapid, as Coley himself stated in 1908: "In successful cases a marked improvement is usually noticed within one or two weeks, often within two or three days."[109]

Time in Memorial

Coley's earliest work was done at New York Hospital and the New York Post-Graduate Hospital. Most of his subsequent work was done at Memorial Hospital, now the clinical arm of Memorial Sloan-Kettering Cancer Center (MSKCC) in New York. His association with this institution began in 1892, when he was appointed attending surgeon of the so-called chronic (or incurable) ward. In the following year, he was made an attending surgeon of the hospital as a whole, a position that he held until he retired in January 1933.

But this hardly tells the whole story. In the same year, 1902, he was elected to the hospital's Board of Managers. He later served for many years on the Executive Committee as well, essentially helping to run the hospital. How did he, as a young man, attain such a high leadership position? No doubt his associates admired his honesty and brilliance. But a big part of the explanation was his fundraising ability. In those days, the hospital seemed to be in a perpetual financial crisis. Everyone seemed to think that treating cancer patients in a specialized hospital was a good idea, but no one had figured out how to make that pay well. Ditto for cancer research. But in 1902, Coley secured the then fabulous sum of $100,000 to establish the Collis P. Huntington Fund for Cancer Research of the General Memorial Hospital. This was the first such cancer research fund in the United States. The grant came from the widow of a celebrated railroad magnate of the day. Luckily, Coley was related by marriage to the Huntington family.

From 1902 to 1909, Coley was secretary of the Fund's Committee in Charge. In 1909, when Dr. Bull himself died of cancer, Coley became chairman of the overall Fund. The research work (which laid the foundations for the present-day Sloan-Kettering Institute) was mainly conducted at Cornell University Medical School under the direction of Bertram H. Buxton, MD. Not surprisingly, Buxton also produced the first Coley's toxins and Martha Tracy's scientific work on the toxins was paid for by the Fund. Later, research was directed by James Ewing (1866-1943) who, as we shall see, was far less favorable to Coley's innovative treatment.

Figure 9 Coley at the time he treated first patient with toxins

But it would be erroneous to think of Coley's treatment as purely a Memorial Hospital phenomenon. It was also widely used by his friends, the Mayo Brothers, at their Clinic in Rochester, MN, as well as by physicians elsewhere in the US, Canada, the United Kingdom, etc. One particularly devoted follower was Henri Matagne, MD, a cancer specialist in Belgium, who—after Coley himself—wrote greatest number of articles on the toxins.[110,111,112,113,114,115]

Mary McClellan Hospital

Because of his many attainments, in 1910, Coley received an honorary Master of Arts degree from Yale University and in 1911 from Harvard University. But, as he contemplated the full sweep of his career (at his retirement banquet at the Waldorf-Astoria in 1933), Coley said that he was especially proud of having helped found the Mary McClellan Hospital in rural Cambridge, New York. Starting in 1919, he served as surgeon-in-chief of that small hospital.[116] Even today, Cambridge is a remote place, almost 200 miles north of New York City, and unserved by any passenger railroad. But on Sunday afternoons, Coley would take the long train ride from Grand Central to Albany's Union Station, and then the 45-mile trip to Cambridge, all to serve the needs of a small upstate community. This was done at considerable personal sacrifice. It meant that "he had only Saturday and Sunday mornings for relaxation," according to his daughter, Helen.[117] He did this mainly out of loyalty to a Yale classmate who had founded the hospital as well as to serve the local rural community.

To all appearances William B. Coley was a blue-blooded member of America's hereditary Eastern Protestant "WASP" Establishment, a life-long Republican, duly listed in the Social Register, friend of the wealthy and the privileged. He was a member of the University Club while his wife was a member of the exclusive Cosmopolitan Club.[118] Yet he was far from typical.

In his award-winning book, *A Commotion in the Blood* (1997), author Stephen Hall grappled with the many contradictions in Coley's "lifelong paradoxes":

> "The boy who milked cows and plowed the fields of his father's Connecticut farm later ministered as personal physician to a rich dowager on one of her European vacations; the boy who walked six miles to Sunday school at the Methodist church in Easton became the man about town who insisted on a chauffeur, in part because he never learned to drive a car…. The boy raised in a culture of thrift and deprivation became the successful surgeon who perpetually lived beyond his means, who dabbled in real estate speculation, and whose financial adventurism was so ill-advised that near the end of his life he could barely pay for medical treatment for his dying wife."[119]

One can only marvel at Coley's energy and determination. For most of his career he worked simultaneously at three or even four jobs, any one of which would have been sufficient for an ordinary doctor's career. He was professor of clinical surgery at Cornell University Medical School; attending surgeon to the General Memorial Hospital; attending surgeon to the Hospital for Ruptured and Crippled; general surgeon to the New York Central Railroad; and Surgeon in Chief at Mary McClellan Hospital in Cambridge, NY.

Yet Coley did all this while secretly suffering from increasingly poor health. As Stephen Hall makes clear, for most of his adult life Coley suffered from acromegaly, a metabolic disease characterized by the overproduction of growth hormones, whose ultimate cause was a benign but inoperable tumor of the pituitary gland. This led to a slow but inexorable development of sausage-like fingers; coarsening and thickening of facial features, especially the nose; thickened skin of the hands and feet; and pain, numbness, tingling and weakness in the hands.[120] All of these developments were appalling for any person but disastrous for a surgeon who depended on his manual dexterity.

In addition, Coley developed stomach ulcers, which (according to his daughter) were exacerbated, if not downright caused, by the aggravation engendered by the bitter opposition of conventional colleagues (such as Ernst A. Codman and James Ewing) to his cancer treatment. In addition, he was severely overworked, especially after he lost most of his savings in the 1929 stock market crash. There was a practical aspect to his busy work schedule…he needed money. Certainly after 1929, his outwardly privileged circumstances masked serious financial problems.

An Ideal Physician

"Dr. Coley is an ideal physician," enthused *Men of Progress* in 1898, "thorough, conscientious, and enthusiastic in his work. Through his writings and successful practice he has gained a reputation achieved by very few physicians of his age."[121]

What makes Coley's personal story of universal interest was its medical dimension. Coley had all the trapping of success. In a sense, his biography reads like that of many other successful and prosperous surgeons, the kind whose portraits grace the walls of every wood paneled medical society in America. Yet, unlike most of them, he put his secure position, as well as his health and hard-earned reputation on the line, as he became increasingly embroiled in a bitter medical controversy. This showed his essential qualities of honesty and courage, and what he himself admired most in a scientist, "doggedness."[122]

Why was Coley so controversial? In part it was that—although ensconced in the cancer establishment of his day—he held heterodox views on the origin of cancer. He persisted in thinking that cancer as a rule was caused by a microorganism, long after most of his colleagues had declared that view hopelessly outdated. He also championed the work of Thomas Joseph Glover, MB, of Toronto, whom most of his colleagues thought of as deluded, if not an outright quack.

But, most significantly, Coley advocated a cancer treatment of his own invention, a mixture of bacterial byproducts for treating cancer. This put him beyond the pale of acceptability for many of his colleagues, and made his name a byword among many right-thinking cancer surgeons, who "knew" that only radical surgery could possibly cure cancer.

Coley's Toxins

Throughout the years, Coley's treatment has gone by several names and this confusion persists into the present-day. Parke-Davis Co., which manufactured a weak form of the product for many years, referred to it as "Erysipelas and Prodigiosus Toxins (Coley)." The American Medical Association (AMA) referred to it in medical Latin as "Toxicum Erysipelatis et Toxicum Bacilli Prodigiosi," and called it "practically a mixed bacterial vaccine." Later on, Coley's daughter often referred to the product as mixed bacterial vaccine, abbreviated MBV.

Coley initially used the term "mixed bacterial toxins of erysipelas and bacillus prodigiosus," – "mixed toxins" for short. Other names were Coley toxins or Coley's fluid. But as early as the 1890s, people logically began to ascribe the treatment to its inventor. The first references to "Coley's toxins" appeared in the mid-to-late 1890s. Some people consider the use of the words "toxins" as derogatory, but in fact most of these early references were sympathetic to the new treatment.[123,124]

Many of Coley's contemporaries regarded his odd-seeming treatment as a radical departure from the kind of medicine that they had been taught in medical school. While this was certainly an era of innovation in medicine, there was nothing with which to compare Coley's toxins. But Coley also had his supporters, even at the highest levels of medicine. As mentioned, the Mayo Brothers of Rochester, Minnesota, were among his active champions and advocated the toxins for cancer patients at their clinic. Henry William Meyerding, MD (1884-1969) was one of those Mayo Clinic doctors who used Coley's toxins in the treatment of bone cancer patients.

Lord Joseph Lister, one of the most famous doctors of his day, was another prominent supporter. Coley recalled in the last year of his life (1936):

> "He did me the great honor of becoming much interested in my early research work in the treatment of inoperable malignant tumors with the toxins of erysipelas and *Bacillus prodigiosus*, and wrote me several long letters about my work in the years 1893 and 1894. In 1895 I had the rare privilege of meeting Lord Lister, and of becoming personally acquainted with him. His interest and sympathetic encouragement were of the greatest help to me in those earlier years when any new method of treating inoperable malignant tumors was looked upon with doubt and skepticism." [125]

But Coley's well-documented advocacy of such an unconventional treatment—including the claim that some advanced patients had actually been cured—brought him into conflict with many other doctors, even at the institution with which he was primarily involved, Memorial Hospital.

It is ironic that William B. Coley, in every other way a doyen of the Establishment, became the leading proponent of a non-conventional cancer therapy in America. There were certainly other irregular cancer practitioners during his time, such as William Koch, MD, PhD of Detroit, Michael J. Scott, MD, of Seattle and the aforementioned Thomas J. Glover, MB, of Toronto. Each of these had his moment of notoriety, but each (whether justifiably or not) was discredited and disappeared from public view. But Coley was different. His treatment successes and failures were scrupulously documented and he stuck to his guns from 1893 until his death in 1936. There followed a small but determined group of people that then preserved knowledge of his achievements with the toxins until the present day.[126]

Growing Opposition to the Treatment

The debate in medicine over whether to encourage or combat fever had gone on for centuries, and swung, with pendulum-like regularity, between these two poles. A major problem for Coley was that at the time that he began extolling the virtues of a fever-inducing treatment in cancer, the mainstream of American medicine was moving in the opposite direction, towards a greater reliance on antipyretic drugs, such as aspirin.

There is no room here to discuss the full scope of the fever controversy. Briefly, in the mid-19th century, the famous German physician, Julius Friedrich Cohnheim (1839-1884), had proclaimed the beneficial effects of fever. He looked on elevated temperature as a manifestation of the "vix medicatrix naturae," or natural healing power of the body, and anticipated a time when physicians would "more and more regard fever, not indeed as a condition free from danger, but as, on the whole, a wise provision of nature."[127]

In America, the tide moved in the opposite direction. "Fever is a complex morbid state characterized by derangement of all the important functions of the body," the *Boston*

Medical and Surgical Journal editorialized in 1890. Fever could cause "respiratory and cardiac symptoms, the loss of appetite and failure of digestion, the muscular prostration, the nervous…disturbances, the urinary derangements, the general impairment of nutrition, etc."[128] The authors looked forward to a new class of anti-fever medications.

Yet Coley's toxins relied in large measure on generating fevers. In fact, by 1896 Coley published 16 papers that described his methods of inducing fever in cancer patients. This went contrary to the mainstream opinion on fevers and indeed on the whole concept of a healing force in nature.

Soon afterwards, editorials began to appear with negative reports, such as one entitled "Failure of the Erysipelas Toxins."[129] In 1895, Robert Abbe, MD, the President of the New York Surgical Society, suggested that Coley's method be tried in different hospitals. It was "a fair proposal, but no enthusiasm was shown," according to Coley's daughter, Helen Coley Nauts. "It is a great pity that such a plan was not carried out, but it is not surprising that a group of surgeons were reluctant to pursue a medical approach to the treatment of cancer."[130] Although Coley was a surgeon, his treatment was purely medical.

On October 16, 1895 Coley read a paper before the New York State Medical Association entitled: "The indications for the non-operative treatment of tumors: The value of toxins." This was reprinted in 1896. "Undoubtedly this report caused more concern among surgeons and a few more negative reports appeared," said Nauts.[131] The editorials that appeared in medical journals in those years were also largely unfavorable and opposition began to mount. Whereas the attitude of the American Medical Association initially was favorable, on Dec. 15, 1894, its journal published a scathing editorial on the "Failure of the Erysipelas Toxins." One of its major charges was that the only a "single investigator," Coley, had reported improvement from use of the toxins.

The charge is rather starting when one realizes that Coley had only begun to use the toxins a year or two earlier. But, even so, this was demonstrably untrue. The month before, Walter B. Johnson, MD, of Paterson, N.J., had reported cure of a very extensive sarcoma of the palate and pharynx, extending to the vocal cords, using toxins prepared for him by Coley. This was published in the *New York Medical Record* on Nov. 15, 1894. As was later reported, this patient remained well at least 7 years.[132]

Coley wrote to the editors of the *Journal of the American Medical Association*. He chided them for having changed their minds "without a more careful inquiry as to the present status of the treatment and a more critical analysis of the evidence presented."[133] He explained that "cultures obtained from any but very virulent strains of erysipelas were of little value, and furthermore that all of my successful cases had been treated with cultures obtained from a fatal case of erysipelas." He concluded:

> "That a few physicians in a very limited number of cases with indifferent preparations of the toxins have failed to obtain good results will not, I am sure,

> have great weight in the minds of the scientific portion of the profession in determining failure or success of this method of treatment of sarcoma."[134]

This was a reasonable position, but was not how most of his colleagues actually acted. Most of them made up their minds based on limited experience, especially after the toxins were condemned by the American Medical Association and its journal.

In March 1895, the New York Surgical Society appointed a committee consisting of three members, Dr. L.A. Stimson, A.G. Gerster and B.T. Curtis to investigate and report on the use of erysipelas toxins in the treatment of cancer.[135] At its meeting on March 25, 1896, the Committee reported as follows:

> "[I]n no case have we found any amelioration which held out a prospect of ultimate cure. We have, on the contrary, observed in some cases that the rate of growth of the disease was much more rapid during the treatment. The treatment also imposes a very severe tax upon the strength of the patient and apparently hastens the cachexia [wasting syndrome, ed.] in most cases."

They also stated that in some cases of apparent cure they doubted the accuracy of the diagnosis.

Their conclusions, therefore, were as follows:

1) That the danger to the patient from this treatment was great.
2) Moreover, that the alleged successes are so few and doubtful in character that the most that can be fairly claimed for the treatment by toxins is that it may offer a very slight chance of amelioration.
3) That valuable time has been lost in operable cases by postponing operation for the sake of giving the method of treatment a trial.
4) Finally, and most important, that if the method is to be resorted to at all it should be confined to the absolutely inoperable cases.[136]

Starting in 1914, these charges were given currency by William Seaman Bainbridge, MD, of the New York Skin and Cancer Hospital, in his popular book, *The Cancer Problem*. Bainbridge reported that the New York Surgical Society had not rescinded the decision rendered by this Committee. Bainbridge then gives a very one-sided view of research on Coley's toxins. For instance, he says of Moullin's confirmatory cases, "Careful study of the history of Moullin's cases at the London Hospital reveals nothing conclusively in favor of the method."

Animal Experiments

Bainbridge showed his bias when he reported on some negative animal studies but then ignored favorable published results. Edward Francis Bashford, MD, director of the Imperial Cancer Research Fund, reputedly told him that he had tested Coley's toxins in mice, with no results. "The fluid is not a cure for cancer," he reportedly said. But Bashford never published these results in a journal. Bainbridge simply cites some personal communication, which was and remains a poor source of documentation.

Meanwhile, Bainbridge ignored favorable animal experiments that were published in the *Journal of the American Medical Association* in 1907, in which two well-respected authors found that both components of Coley's toxins had anticancer activity in lymphosarcoma in dogs. Silas Palmer Beebe, MD, PhD, and Martha Tracy, MD, reported that *Bacillus prodigiosus* (i.e., *Serratia marcescens*) alone and *Streptococcus pyogenes* alone as well as the mixture of these two (Coley's toxins) caused regression of this tumor in dogs. All the tumors in question disappeared in five weeks. Intra-tumoral injections caused the most rapid regressions. [137]

Beebe had a PhD from Yale (1904) and a medical degree from Cornell (1908). He worked as a physiological chemist for the Huntington Fund for Cancer Research. Martha Tracy was another worker at the Fund who later produced some of the finest versions of Coley's toxins ever made. In their conclusions, Beebe and Tracy stated: "Though the action is chiefly local, it is at the same time something more than this for it was repeatedly observed that tumors at a distance from the site of injection undergo regression simultaneously while in one dog all the injections were given remote from the tumor."[138]

Bainbridge cites other irrelevant papers by Beebe and Tracy but fails to cite their *JAMA* article. Instead, he only cites a negative unpublished report by Bashford. In my opinion, this revealed a deep bias against Coley, and against non-conventional treatments in general. [xiv]

Bainbridge also tells the story of Sir Frederick Eve's failure to achieve results in 5 cases of periosteal sarcoma at London Hospital. Many of these British doctors started out as enthusiasts, but became disillusioned or even opposed to the treatment because of this experience. It is perfectly understandable why skepticism grew under the circumstances. It was based on such frequently repeated stories of failure that Sir Frederic Eve, MD (1853-1916), a senior surgeon at London Hospital, denigrated the use of Coley's toxins in print.

[xiv] Bainbridge and Bashford were also determined foes of the pancreatic enzyme treatment of John Beard, DSc of Edinburgh. The Imperial Cancer Research Fund doomed that treatment, also without ever publishing their results in a peer-reviewed journal, while Bainbridge's *The Cancer Problem* in effect wrote the obituary on this treatment. See: Moss, RW. The Life and Times of John Beard, DSc (1858-1924). *Int Cancer Ther.* 2008;7:229-251.

Photograph by] SIR FREDERIC EVE, KT. [Elliott and Fry.

Figure 10 Sir Frederic Eve (1853-1916)

Source: British Medical Journal, Dec. 23, 1916 (via PubMed Central)

Eve was highly regarded and his article in the *Lancet* may have kept Coley's treatment from receiving further serious consideration in Britain and consequently in many other places around the world.[139] Eve admitted that Coley's toxins sometimes actually destroyed tumors. "It is undeniable," he said, "that a certain number of sarcomata disappear under, or are profoundly influenced by the toxins." However, "we have no means of determining what proportion this is to the total number treated. I suspect it is only a small proportion.... I would not recommend Coley's fluid in any case of operable sarcoma, nor would I recommend it as a prophylactic agent against recurrence."[140]

Bainbridge's argument also included the fact that he himself had observed 30 cases in which Coley's fluid "was used without permanent benefit." Note the word "permanent," as if there were any other treatment in the world—other than surgery for localized tumors—that had a *permanent* benefit in cancer. He also claimed to have seen "hundreds of cases" at his own institution, the Skin and Cancer Hospital of New York (long ago absorbed into New York University) "in which the mixed toxins had already been employed elsewhere without success."[141] This is another fallacious argument that is often used against competitive treatments—"I see their failures." The problem, of course, is that each doctor necessarily sees a lot of other doctors' failures. By the nature of things, they don't see many of their successes (since such people generally have no need to seek further assistance).

In addition, Bainbridge said he had surveyed "5,000 representative medical men throughout the civilized world," who, with few exceptions, were unfavorably inclined towards the treatment. I find this statement preposterous. Assuming that it is true, relatively few of these surveyed doctors could have had first hand experience treating cancer with the toxins. The majority was probably echoing the prevalent prejudices of their time.

If you read this critique carefully you will see that Bainbridge knew how to choose his words very carefully. Notice that he and his colleagues do not deny outright that Coley's toxins could be effective. "A certain number of sarcomata disappear under...the toxins," to quote Dr. Eve. James Ewing similarly wrote in *Neoplastic Diseases:* "[I]n some recoveries from endothelioma of the bone, there is substantial evidence that [Coley's] toxins played an essential part" (see below).

So sometimes the treatment worked. Coley's opponents therefore directed their fire at the fact that results with the toxins were variable, and that tumors in some (many?) cases tended to recur. They also stated—without the slightest bit of documentation—that the adverse effects of the treatment were more than just fever and chills, but that the toxins could exacerbate cachexia (the wasting syndrome), and thereby lead to the premature death of these patients. To complete their bill of charges, they claimed, again without any evidence, that patients resorted to Coley's treatment instead of submitting first to 'curative' surgery. The exact opposite seems to have been the case in the great majority of Coley's patients.

In the 21st century there are three highly developed modes of cancer treatment (surgery, radiotherapy and chemotherapy) as well as numerous experimental treatments available through clinical trials. It is therefore difficult for us to put ourselves back in the mindset of 100 years ago, at the time Bainbridge wrote. At that time, besides some 'alternative' treatments such as Coley's toxins or pancreatic enzymes, there were virtually no cancer treatments other than radical surgery. And surgery itself was often a dangerous and uncertain business. For instance, at that time the death rate from cystectomy (surgical removal of the bladder) was at least 50 percent.[142]

Chemotherapy was half a century away and radiotherapy was still in its infancy. In most cases, as Coley himself showed, the final results with radiation were negative, since recurrences or continued growth of the tumor, followed by death, was the invariable outcome.[143]

So it seems remarkable that Bainbridge and his colleagues would admit the essential validity of Coley's observation—sometimes the toxins worked!—but then consigned the treatment to the garbage can because of inevitable uncertainties about the frequency of response or their permanence. Bainbridge also does not seem to have considered that Coley's toxins, because they were based on the cultivation of living organisms, might be of variable quality, and that poor results (as in the case of Sir Frederic Eve) might have been the result—as Coley repeatedly tried to point out—of low standards by some of the manufacturers (including Parke-Davis Co.).

In 1914, Torr Wagner Harmer, MD, a Chestnut Hill, Boston surgeon, also published a report on 134 microscopically proven cases treated by Coley's toxins. He concluded that they were of value in certain cases of inoperable sarcoma. He also noted that in a small number of cases they produced striking relief of pain. But overall his conclusions were, like Bainbridge's, rather negative.[144]

It is significant that many of Coley's supporters in England and in the United States urged him to limit his use of the toxins to sarcomas. This was because the early experiences with his weaker preparations in advanced carcinomas or melanomas had not proved successful. To this day there remains a school of thought that Coley's toxins are only effective in sarcomas, and in no other kinds of cancer. But reports of benefit in non-sarcoma cases led Coley to try his method in many other types of malignancy, including testicular cancer, breast cancer, lymphoma and Hodgkin's disease, malignant melanoma, multiple myeloma, neuroblastoma, colon cancer, and renal cancer, "with a great many remarkable results," according to Helen Coley Nauts. She noted that "other surgeons had successfully treated metastatic cervical carcinoma, extensive metastatic breast carcinoma, giant cell tumor of vertebrae, recurrent malignant melanoma, and metastatic ovarian." [145]

Opposition of James Ewing

I have spoken about the opposition of Bainbridge, who was a major figure in New York cancer circles. But Coley had an even bigger problem. He also faced the opposition of James Ewing, MD (1866-1943), who came to Memorial in the same year as Coley (1902), became its chief pathologist and, starting in 1913, medical director of the nation's first cancer hospital. In the period between the two World Wars, Ewing was the public face of cancer research in America.

Figure 11 James Ewing, Jan. 12, 1931

One sometimes sees Ewing described as Coley's "boss" at Memorial.[146] But this overstates the case, as Memorial surgeons of the time were fairly autonomous free agents.

But as an increasingly dominant figure at Memorial and as a public spokesperson, Ewing played a major role in the fate of Coley's toxins. A word about his life and career is therefore in order.

James S. Ewing was born on Christmas Day, 1866, the third of five children of Judge Thomas and Julia Hufnagel Ewing. He came from a wealthy and influential Pennsylvania family. In 1880, at age 14, he was afflicted with a bone infection called osteomyelitis of the femur, which confined him to bed for the next two years. During this period of recuperation he entered various competitions, in one of which he won a microscope. These two factors—bone disease and microscopy—were to determine his choice of career. After recovering, he studied Classics with Henry Gibbons at the University of Pennsylvania and mathematics with C.B. Wood. He then attended Amherst College in Massachusetts and graduated with a bachelor's degree in 1888. He attended the College of Physicians and Surgeons of New York from 1888 to 1891 and received his MD from Columbia University in 1891. (He also received an MA from Amherst in 1892.)[147]

Ewing studied medicine in Vienna, and after his return was, successively, tutor, fellow and instructor at Columbia University.[148] In 1899, at age 33, Ewing became the first professor of clinical pathology at the Medical College of Cornell University. In 1900 he married Catherine Halsted and their son James, Jr., was born in 1902. In 1902, Ewing began to do cancer research at the Loomis Laboratory for Research in Experimental Pathology, which was affiliated with Memorial Hospital (and where Coley's toxins were produced).

In addition to his scientific achievements, Ewing was said to possess "a keen artistic temperament,"[149] and this increased his suffering from repeated personal tragedies and ill health. The greatest tragedy was the death, on July 26, 1903, of his young wife, Catherine, age 25, and their unborn second child. This left Ewing responsible for the care of their one-year old child, James, Jr.[150] According to one biographer, this loss "evoked reclusive and eccentric tendencies" in Ewing's personality. From US census references I infer that he later remarried, had a son named John, and lived with his second wife in Westhampton, Long Island. But some of the census records give his domicile as a boarding house in Manhattan, where he probably stayed when he was working at Memorial. He had at least two grandchildren, but there remains little information on his personal life.

Ewing and Coley's careers, and lives, interwove in various complex ways. They were colleagues in the goverance of Memorial Hospital. In 1907, they were among the cofounders of the American Association for Cancer Research (AACR), which soon became an influential organization in the cancer field. Ewing was interested in public education and co-founded the American Society for the Control of Cancer, which is now the American Cancer Society.

Cancer research did not receive appreciable governmental funding until the late 1930s and so scientists invariably had to seek support from wealthy individuals, such as the Huntington railroad family. One of Ewing's most valuable contacts of those days was

James Douglas, a mining engineer with whom he founded the National Radium Institute in 1913.

In 1913 Ewing was also elected president of the Medical Board of the General Memorial Hospital, thereby in a sense becoming Coley's superior. He was also first director of research and director of Memorial Hospital (from 1931 to 1939), a position that laid the groundwork for the later formation of Sloan-Kettering Institute (1945). Ewing is generally credited with the intellectual parentage of the "modern comprehensive cancer center," a concept that combines research and treatment ("bench and bedside") in a single institution. He retired in 1939 and died of bladder cancer in 1943 at the age of 76. Over 1,000 people attended his last rites.

One little appreciated fact about Ewing is that for most of his life he suffered from trigeminal neuralgia ("tic douloureux"), a debilitating chronic disease that has been called "the most terrible pain known to man."[151] This affliction greatly exacerbated his already dour personality and diminished his ability to work, although, by our standards, his output was prodigious. His textbook, *Neoplastic Diseases*, which initially took him ten years to write, went through four revisions (1919, 1922, 1928 and 1942) and is still considered "the cornerstone of modern oncology."[152] It was in this book, incidentally, that he first expounded the discovery of the bone cancer that is named after him, Ewing's sarcoma.

In his approach to the cancer problem Ewing was a cautious reformer while Coley was a revolutionary. Ewing was naturally skeptical of "cures" and the most he hoped for was slow but steady progress. "I do not look for any startling advances or sensational discoveries," he said in 1913, "since it is much more likely that a steady reduction in the mortality from cancer will come chiefly from a large number of separate factors, of which the most significant appear to be increased control of the conditions leading to cancer, more general recognition of the preliminary stages of the disease, early diagnosis, and treatment of the established disease."[153]

As one might expect, Coley and Ewing had a complex relationship. They were considered two of the four leading American bone cancer specialists.[xv] [154] But there was a crucial difference. *Ewing did not treat cancer.* He was an expert pathologist, whose ability to detect and diagnose cancers, especially bone cancers, was considered unparalleled. Coley, by contrast, was one of the people charged with the responsibility of treating many of those same cases. He was considered one of the greatest surgeons of his day.

Surgeons and pathologists are mutually dependent but often do not always see eye to eye. A contemporary commentator likened them to two sides in a cricket match.[155] Ewing was highly competitive and believed that some of Coley's "successes" were what were politely called "indeterminate cases." In other words, he would not certify that these patients actually had cancer, thereby undermining some of Coley's claims. To complicate

[xv] The other two were Joseph Colt Bloodgood, MD of Johns Hopkins School of Medicine and Frank Burr Mallory, MD of Harvard Medical School.

matters, Coley claimed some of his best results in *Ewing's sarcoma*, a diagnosis in which, needless to say, Ewing's own assertions were held supreme, not least by Ewing himself.[156]

There were other important differences between the two men. One had to do with the etiology of cancer. In addition to his advocacy of the toxins, Coley stood apart from his colleagues because he persisted in believing that the cause of cancer would ultimately be found in microbes. This had been a popular position when Coley and Ewing were in medical school, but had gradually lost favor in the intervening years.[157] Yet Coley saw the microbial theory as fundamental to the eventual solution of the cancer puzzle:

> "To those of us who believe that cancer is of infectious nature, and due to some microbic organism or group of closely allied microorganisms, the discovery of such organisms offers the only real hope of solving this greatest of medical problems."[158]

Ewing, in turn, became an acerbic critic of the microbial theory. Belief in this theory, he said, had peaked around 1895 but then had quickly lost support in the medical world. By 1919, he wrote in his authoritative *Neoplastic Diseases*, "Few competent observers consider it as a possible explanation for the unknown element in blastomatosis," i.e., tumor formation.[159] So Coley remained in a minority—a shrinking minority—on this point, although history has justified his position somewhat. The microbial theory has gained adherents in recent decades, a point made by Prof. David Hess in his book, *Can Microbes Cause Cancer*?

We needn't delve too deeply into this controversy. Suffice it to note that the National Cancer Institute Web site (www.cancer.gov), which reflects the present-day consensus, now contains the following statement: "Research has indicated that infection with *H. pylori* [a type of bacteria, ed.] increases the risk of gastric cancer, gastric mucosa-associated lymphoid tissue (MALT) lymphoma, and possibly pancreatic cancer."[160] Similarly, there are links of human papilloma virus with cervical cancer, of human immunodeficiency virus (HIV) with Kaposi's sarcoma and of Epstein-Barr virus (EBV) with Hodgkin's disease, Burkitt's lymphoma and nasopharyngeal cancer. Such statements, no longer controversial, would probably have stupefied Ewing. So in a sense the tide on this question has turned somewhat in Coley's favor.

Hess suggests that Coley's interest in the microbial etiology of cancer may have been a very significant factor in the failure of his own treatment to gain acceptance. "It may have even heightened [James] Ewing's dislike for the treatment," Hess has said. "After all, couldn't radiation and Coley's toxins be combined as therapies? Coley's interest in the bacterial etiology was not a politically astute choice, as his daughter (I think) recognized."[161]

Even more problematic was Coley's championing of an unorthodox cancer "serum" proposed by Thomas Joseph Glover, M.B., of Toronto. There is no space here to trace the Glover controversy. Basically, Glover claimed to have recovered a pleomorphic (everchanging) organism from the blood of cancer patients, and to have produced an antiserum against it, recoverable from the blood serum of horses. As Hess shows in *Can Bacteria Cause Cancer?*, Coley's defense of Glover—even after the latter had been thoroughly trounced by the medical establishment—became the source of considerable embarrassment and grief for Coley and further tarnished his reputation. I think it speaks volumes about his integrity, however, that he was willing to support an unconventional theory at great personal cost.

Ewing's Opposition to Biopsies

Ewing was not as thoroughly orthodox as many people thought. He also held some very unconventional views on cancer, which also brought him into conflict with Coley. Rather startlingly, he was strenuously opposed to the use of biopsies, which he felt could release or track cancer cells from the disrupted tumor. Most of the medical profession by the 1920s and early 1930s had become convinced of the necessity of tumor sampling before surgery. Coley held a more conventional view on their safety and utility. But this was a source of tension.

Coley and Ewing also held irreconcilable views on the usefulness of radiation therapy. Coley was actually one of the first doctors in New York City to routinely use Roentgen's "new kind of light" in his practice. As early as 1901 he persuaded Memorial Hospital to procure the new-fangled x-ray machine, which was paid for by one of his wealthy patients. He read his first report on the use of radiation before the American Surgical Association in June 1901 and was also a leading member of the Roentgen Ray Society.

But, as time went by, he became increasingly skeptical of the long-term benefit of radiation therapy. This put him into conflict with Ewing and others, who became increasingly vocal advocates of this mode of treatment. In fact, Ewing became such a strong advocate that he once said that every person who experienced chronic bone pain should be treated as if he had bone cancer and given prolonged radiation treatment, without the need to even be given a biopsy![162] It was an eccentric position that, happily, never was adopted by the rest of the surgical profession.

A clash between the stubbornly persistent Coley and the authoritarian Ewing now seemed inevitable. The conflict was not helped by the fact that each man was simultaneously suffering from an intractable and painful illness, acromegaly in Coley's case, trigeminal neuralgia in Ewing's. Each of these diseases, in their own way, made pleasant social interactions problematical. According to *A Commotion in the Blood*, Ewing's negative attitude towards Coley first surfaced in 1912, when he conspicuously failed to accept the validity of Coley's data in his article "The Treatment of Cancer on Biological Principles," published in the *New York Medical Journal.*[163] If there were ever an opportunity for Ewing to praise the toxins, this was it. At that time Coley already had recorded "200 durable cures in inoperable and supposedly hopeless cancer cases,"[164] so the omission was pointed.

In private, Ewing made no secret of his antipathy towards Coley's toxins and even towards Coley personally. We have to remember that this was an era in which doctors never criticized one another publicly, least of all their colleagues at the same hospital. So, in public, Ewing maintained the pretense of a tentatively favorable attitude. In *Neoplastic Diseases*, for instance, he wrote, "[I]n some recoveries from endothelioma of the bone, there is substantial evidence that [Coley's] toxins played an essential part."[165]

At Coley's January 1933 retirement banquet at the Waldorf-Astoria, Ewing waxed enthusiastic over Coley's treatment, striking the right tone for the occasion. "In certain types [of cancer, ed.]," he said, Coley has "effected miraculous cures in substantial numbers all over the world."[166] These were fine words for the occasion. But in private he was negative and even seemed determined to ruin Coley's reputation. In a July 1917 letter, for instance, he wrote: "You will be discredited in the end."[167] These were fighting words, which belied his amicable public attitude towards his colleague.

Reception of Coley in his Lifetime

In early 20th century America, there were frequent controversies over the proper treatment of cancer. In general, medical treatments were given short shrift by the surgical establishment. We needn't mention the treatment afforded unconventional pioneers of the day, such as Koch, Glover and Scott. Surgeons of the time looked on cancer chemotherapy as little better than quackery.[168] Even Peyton Rous, MD (1879-1970), of Rockefeller University, was scorned when he proposed the viral etiology of certain animal cancers in 1909. The value of this work was only recognized half a century later, and Rous was belatedly awarded a Nobel Prize in 1966.[169] The reception of Coley's toxins in his own lifetime was mixed. The reason it wasn't more negative, I feel, was because Coley held such a high social and medical position in New York and so could not be dismissed out of hand, the way other cancer mavericks often were.

Amid a background of suspicion, Coley achieved a certain measure of acceptance, even celebrity. In 1897, at the age of 36, he was chosen a fellow of the American Surgical Association, the youngest ever then elected; he was a fellow of the Southern Surgical and Gynecological Association; the Pathology Society; the New York State Medical Association; the American Academy of Medicine; and the New England Society of Medicine.[170] He was also the author of the section on cancer for the textbook *Twentieth Century Practice of Medicine.*

As chief of the Bone Service at Memorial Hospital, which he founded around 1915, Coley built up a national reputation and probably treated more such cases than any other doctor in the US.[171] He also had an international reputation. In July 1909, he was invited to give an address on his cancer treatment to the Surgical Section of the Royal Society of Medicine in London. He was made an honorary fellow of the Royal College of Surgeons in London, one of five living American surgeon to be so honored. (The others were Drs. Harvey Cushing, George W. Crile, and Charles H. and William J. Mayo.) In 1911, he was invited by the President of France to give an address to the French Congress of Surgeons in Paris.[172]

The New York Times praised his work. "Erysipelas Germs As Cure for Cancer," it announced in 1908. "Dr. Coley's Remedy of Mixed Toxins Makes One Disease Cast Out the Other. MANY CASES CURED HERE. Physician Has Used the Cure for 15 Years and Treated 430 Cases—Probably 150 Sure Cures."[173] The article stated: "It came out yesterday that nearly 100 cases of that supposedly incurable disease have been cured in this city [New York, ed.] during the last few years…." I think this article unintentionally speaks volumes about Coley's modesty; he had, after all, been successfully treating cancer patients since the early 1890s. *The Times* also said that Coley "has more cancer cures to his credit than any other surgeon in the world." Yet his 'hometown paper' only learned of the work 15 years later.

Coley never sought public acclaim, but a 1933 *Time* magazine article, coinciding with his retirement, finally brought him to national attention. According to *Time*, the microbial products in Coley's fluid "have cured an appreciable number of people dying of cancer, and together they have in numerous instances prevented the spread of cancer to remote parts of the victim's body." It was a remarkable statement, coming from one of the dominant "establishment" media outlets of its day. Time also pointed to the softening of the attitude of perhaps the key group, the American Medical Association (AMA), towards Coley's work.

In 1894, the *Journal of the American Medical Association (JAMA)* had issued a sweeping condemnation of the use of the toxins:

> "There is no longer much question of the entire failure of the toxin injections, as a cure for sarcomata and malignant growths. During the last six months the alleged remedy has been faithfully tried by many surgeons, but so far not a single well-authenticated case of recovery has been reported."[174]

In 1909, the American Medical Association (AMA) had relented a bit and hesitantly approved Coley's fluid as a "New and Nonofficial Remedy." But in 1933 "with much less hesitancy," *Time* said, the AMA reaffirmed its approval. This time it did so enthusiastically. In a three-column report and a two-column editorial, *JAMA* actually recommended that Coley's toxins be used in certain situations.[175] The citation read as follows:

> "It appears, that undoubtedly the combined toxins of erysipelas and prodigiosus [*Serratia marcescens*, ed.] may sometimes play a significant role in preventing or retarding malignant recurrence or metastases; occasionally they may be curative in hopelessly inoperable neoplasms; The Council has, for these reasons, retained Erysipelas and Prodigiosus Toxins Coley in New and Nonofficial Remedies, with a view towards facilitating further studies with the product."

Coley himself was the first to admit that he did not understand the mechanism of action and that it certainly was not effective in every case. "Nonetheless the fluid works," said *Time*. "Not a cure-all, it has proved effective in a large percentage of inoperable bone cancers."[176] It was an extraordinary homage to an unconventional cancer treatment and arguably the high water mark of acceptability for Coley's toxins in the US—indeed for any "alternative" therapy ever proposed.

Yet Coley's toxins got no further. There were a number of reasons for this. Coley himself was old and ailing at the time. His acromegaly, which is apparent even in early pictures, had become alarmingly pronounced. Although officially retired, his finances, depleted by the Depression, kept him working. At this point, he started once again to use intratumoral injections of the toxins, repeating the success he had achieved with this mode of administration early in his career. Most commentators believe that Coley's failure to use intratumoral injections in the intervening years limited the effectiveness of the treatment.

William B. Coley died on April 16, 1936, at the Hospital for the Ruptured and Crippled, after emergency surgery for just the sort of an intestinal infarction that he himself routinely treated there. He was 74 years of age. More than 500 medical and civic leaders attended his funeral services—to the strains of Dvorak's *New World Symphony*—at the Brick Presbyterian Church on East 92nd Street. Although the announcement called for no flowers, his friend John D. Rockefeller, Sr., sent a large cluster of roses.[177] Ironically one of the pallbearers was James Ewing. Dr. Coley's body was then taken to Sharon, Connecticut, where he was interred in a cemetery near his beloved summer home, Okeden.[178]

Bradley Coley, MD (1892-1961)

In 1933, upon William B. Coley's retirement from Memorial Hospital, his son Bradley who was his father's partner and colleague, took over his position as head of the Bone Service as well. He also assumed his father's position with the New York Central Railroad.

Bradley was the logical person to bring his father's work to a successful conclusion. However, his position vis-à-vis his father was conflicted (as often happens with the sons of great fathers). A scrupulous scientist, with a careful and logical mind, he nevertheless lacked the passionate attachment to an overarching idea that characterized his father and, later, his sister. That said, we must appreciate that when Bradley took over the task of championing the toxins there was growing dogmatism concerning the uselessness of all non-conventional cancer treatments. Radiation and chemotherapy were in the ascendancy, competing for the same position and patients occupied by Coley's toxins.

Figure 12 Bradley Coley, MD

Source: National Library of Medicine

Bradley Coley was born in New York City on Dec. 23, 1892. Like his father, he graduated from Yale University but received his medical degree from Columbia College of Physicians and Surgeons in 1919. He and his wife Phyllis (née Macdonnell) of Vancouver, had three sons. Bradley Coley and his family lived at 70 East 96th Street. For the most part, he followed in his father's footsteps. Bradley also served as surgeon for the powerful New York Central Railway. In 1921, he joined his father's practice at 5 West 50th St., New York, and was elected a member of the New York and New England Railroad Surgeon's Association, where he remained until his retirement nearly 40 years later. Bradley joined the staff of Memorial Hospital in 1921 and worked in the Bone Service, taking over its chairmanship upon his father's retirement. He then led that department until he reached the statutory retirement age in 1959. Like his father, he never really retired but remained in private practice until his own death on June 1, 1961 at the age of 68.

Bradley Coley is best remembered for pioneering an operation for removal of the hip joint in bone cancer, which became a standard procedure. In 1949, he published a well regarded textbook, *Neoplasms of Bone*, that was justifiably appreciated in its day.[179] Like his father, he was a professor at Cornell University Medical School and on the staff of the Hospital for Special Surgery. During World War II, he served in the US Army in the Philippines and later founded the Amputees Alliance, a self-help group for crippled persons.

Bradley Coley collaborated with his father, and later his sister, on articles endorsing the use of the toxins. He also included a chapter on the toxins in *Neoplasms of Bone*, the first such exposition in a medical textbook. This chapter, like the rest of the book, is scholarly and well written. But Bradley Coley was very cautious in his assessment of his father's notorious treatment. Obviously, he had to protect himself from the charge that he was only promoting the treatment because of its association with his father.

Reading this chapter, one has the impression that results in Bradley Coley's day were not as good as in William B. Coley's day, especially the early days of the treatment. Bradley ascribes two reasons for this. First, by the 1940s, doctors only had access to the Parke-

Davis commercial formulas, which were markedly inferior to the earlier Buxton and Tracy formulas that were produced for Coley's personal use (especially the one later designated Buxton VI, Tracy X and Tracy XI). Also, by the late 1940s, patients were more likely to undergo surgery and radiotherapy in addition to the toxins than they were in William B. Coley's day. Starting in the 1950s, chemotherapy and other systemic treatments further complicated the picture. Results after such toxic treatments were probably less dramatic than in "virgin" patients. Bradley Coley also had to be scrupulously careful not to attribute successes to Coley's toxins that may also have been due to the concomitant use of other possibly curative treatments. And such concomitant use was becoming increasingly frequent.

Bradley was a good scholar, with a detailed knowledge of the history of both erysipelas and cancer. He had a reputation as an excellent surgeon as well. But, judging from his writing and from personal comments that his sister made to me, he lacked the passion that his father brought to the struggle for the mixed bacterial vaccine. I am extrapolating, but perhaps some of his caution was a reaction to his younger sister's passionate advocacy of their father's ideas.

We must also remember that Bradley Coley lived through a time of diminishing acceptance for new and challenging ideas in the cancer field. When the senior Coley practiced there were only two treatments for cancer, surgery and radiation, and the latter was not yet on a firm footing. But by the time Bradley retired, chemotherapy had emerged, at least as a concept, and sometimes as a practice. So the leaders of Memorial Sloan-Kettering, led by Cornelius P. "Dusty" Rhoads, were increasingly enamored of chemotherapy and intolerant of other competing therapies.[180]

According to Mrs. Nauts, "Bradley Coley did not discontinue the use of the toxins in the Bone Tumor Department until he was forced to by Dr. 'Dusty' Rhoads, about 1955. He [Bradley Coley] and Dr. [Norman L.] Higinbotham were furious as they were treating several patients at the time who were doing well. He and Higinbotham between them cured quite a few cases in the 40's and 50's."[181]

Parke-Davis ceased producing Coley's toxins in the 1950s. The critical turning point came in 1963 when the Food and Drug Administration (FDA)—over the strenuous objections of Helen Coley Nauts and her followers—ruled Coley's treatment a "new drug" and therefore illegal to administer in the US. FDA refused to 'grandfather' the toxins into existing practice, as a medicine that had been accepted for 70 years, the way it did with aspirin and other long established products. Whatever he thought in private, Bradley Coley did little publicly to resist these changes. Judging from his writing, his attitudes were nuanced and complex, but this was a time that called for decisive public opposition. Fewer qualms inhibited his younger sister, Helen, who was not restrained by a medical degree from fully speaking her mind.

Helen Coley Nauts

Detailed knowledge, much less the continued exploration of Coley's toxins would probably have been lost forever had it not been for the Herculean efforts of William B. Coley's daughter, Helen Coley Nauts. In the story of the mixed bacterial vaccine, her place is just below that of her illustrious father. In a way, her story is more even more extraordinary.

Helen Lancaster Coley was born in 1907 and graduated from Miss Porter's School in Farmington, Conn., in 1925. (This elite boarding school also educated the young Jacqueline Kennedy.) In September 1928, Helen married a banker from Tennessee named William Boone Nauts, Jr.[182] Her father's younger brother, the Rev. Edward Huntington Coley, officiated at her wedding, which was held at Christ Church, in the Coley's summer hometown of Sharon, Conn.

The Nautses lived on East 79th St. in Manhattan and in 1929 moved to the Long Island suburb of Great Neck, NY. After her parents' deaths (both died in 1936), Helen inherited Okeden.[183] There, in the barn, she found her father's voluminous papers, including a stash of 15,000 letters and other memorabilia. She started organizing these papers with the thought of possibly writing his biography. The more she read, however, the more she learned and marveled at the efficacy of his unusual bacterial treatment of cancer.[184]

Figure 13 Coley with daughter Helen, c. 1911

Source: Cancer Research Institute

Although Helen was not a scientist, but a housewife with only a high school diploma, she—against all expectations—ultimately became recognized as an outstanding authority on the immunological dimensions of cancer and a skilled advocate for her father's cause. She engineered the resuscitation of his treatment and in doing so helped to create the modern field of cancer immunotherapy. In the process, she developed the skills, knowledge and vocabulary of a professional medical researcher. For instance, here is a description of her interview of Gregory Shwartzman, MD, of Mt. Sinai Hospital, New York. She later recalled:

> "I had read Shwartzman's book on the reaction of tumors to bacteria, and had prepared 13 pages of questions based on it. During the interview, I took 80 pages of notes. Shwartzman couldn't believe anyone could be that thorough. But because I had no medical education, I had to be that way."[185]

In 1946, she was first author on a comprehensive review of her father's work, published in *Cancer Research* (flagship of the American Association for Cancer Research). In 1953, Mrs. Nauts, with the help of financier Oliver R. Grace (1912-1992), founded the New York Cancer Research Institute to promote research into Coley's methods. (The name was later shortened to the Cancer Research Institute, or CRI.)

The Monographs

Mrs. Naut's most enduring legacy (aside from the Cancer Research Institute) was writing and/or editing a series of 18 comprehensive monographs (as well as various papers and lectures) on her father's results in various kinds of cancer. These are too voluminous to summarize here. But I would like to present the outcomes that she found in one class of cancer, soft tissue sarcoma. These were presented in a 219-page monograph published in 1975. [186]

This monograph summarized the outcomes in 186 cases. Of these, 49 were operable cases, in which the toxins were given as an adjunct to surgery. The rest were deemed inoperable at the time that they received the toxins (see Table 1).

Table 1 Successes and Failures in Soft Tissue Sarcomas

	No.	Percent success
Inoperable successes	70	51.1%
Inoperable failures	67	48.9%
Operable successes	35	71.4%
Operable	14	28.6%

failures		

Each of the cases was then presented in abstract form. One can imagine the type of detective work that was required to establish the facts in such a case, including determining both the outcome and cause of death. Multiply that effort by a thousand and you get some idea of Mrs. Naut's prodigious output.

Although she eventually won the National Institute of Social Sciences' Gold Medal for Distinguished Service to Humanity (the only previous female winner was Marie Curie.), I do not think that Mrs. Nauts's scientific contribution have been adequately recognized. Her career was proof that sometimes the seminal contributions to a field come not from its professionals but from talented outsiders who are, perhaps, better able to see the forest for the trees.

Lloyd J. Old, MD

Throughout most of her career, Mrs. Nauts was greatly helped by her association with an influential group of scientists at Memorial Sloan-Kettering. Chief among these was Lloyd John Old, MD, Director of the Ludwig Institute for Cancer Research. Old was born in San Francisco on Sept. 23, 1933. He graduated from the University of California, Berkeley with a bachelor's degree in biology and attained his medical degree from the University of California School of Medicine San Francisco (UCSF). In the late 1950s he began his half-century association with Memorial Sloan-Kettering in New York. In 1959—fresh out of medical school—Old investigated the effect of the tuberculosis vaccine, *Bacillus Calmette-Guerin (BCG)* on cancer.[187] This led to the use of intracavitary BCG as a postoperative treatment of bladder cancer.[188] This was the first approved use of immunotherapy in contemporary oncology. He went on to make numerous important discoveries, including tumor necrosis factor (TNF) and the CD family of antigens.[189]

In 1971, Lloyd Old became director of the Cancer Research Institute's Scientific Advisory Council. A few years later he was made vice president of Sloan-Kettering Institute. Old called Mrs. Nauts "one of the leading scholars in the field of cancer immunology." Because of her tireless efforts in promoting his work, he said, William B. Coley was now widely recognized as the father of cancer immunotherapy.[190]

Aside from the Coleys, Dr. Old did as much any other individual to preserve and extend the bacterial treatment of cancer. One of their achievements was to get Coley and his methods removed from the American Cancer Society's *Unproven Methods of Cancer Treatment* "quack" list.[191] That notorious ACS article had begun:

> "After careful study of the literature and other information available to it, and in view of the length of time which these toxins have been under investigation without any scientifically acceptable favorable reports, the American Cancer

> Society has found no evidence that treatment with Coley's mixed toxins results in any objective benefit in the treatment of cancer in human beings."
> The formulaic phrase "after careful study" hid a bitter irony. The ACS statement on each of the unproven methods was a superficial and sloppy compilation of research and hearsay evidence. In fitting with the McCarthyite temper of the times, the ACS's condemnation was a kind of Procrustean bed, onto which each method had to fit, willy-nilly, so that it appeared to be worthless quackery. Most of the treatments were never in fact tested and in some cases had been tested and had shown some positive data.

In the 1990s, the ACS eventually withdrew these statements without comment, and we might add without apology to the various scientists whose lives and careers had been devastated by their blind opposition.

Because of the determination of Mrs. Nauts, Coley's toxins were eventually removed from the "quack list" before the general demise of the ACS project. In 1975, the tide began to turn in favor of the immune treatment of cancer. With the discovery of tumor necrosis factor (TNF) by Lloyd Old and his colleagues at Sloan-Kettering, and the emergence of other cytokines, a new era of "fourth modality" of cancer treatment—immunotherapy—seemed to be at hand.[192] Good was lauded in a cover story in *Time* magazine. In this context, and under pressure from Sloan-Kettering scientists, the ACS reversed its previous opinion of Coley's toxins. It also removed staphage lysate, Lincoln bacteriophage and hyperthermia from the list as well. In 1975, this event was celebrated at a black tie gala at New York's Pierre Hotel. It was a publicity coup for the intrepid Mrs. Nauts.

Not long afterwards ACS spent $2 million to purchase a cytokine, interferon-alfa, for the treatment of advanced cancer.[193] At this time, the National Cancer Institute (NCI) also invested huge resources into TNF and interleukin-2 trials.

The highlight of all this activity was the extraordinary series of monographs, in which Mrs. Nauts, often with her collaborators, George A. Fowler, MD and Louis Pelner, MD, analyzed the outcome of Coley's treatment on a case-by-case basis, in such cancers as sarcoma (1959),[194] colorectal cancer (1969),[195] malignant melanoma (1968),[196], testicular cancer (1968),[197] lymphosarcoma (an old name for non-Hodgkin's lymphoma, 1969),[198] and neuroblastoma (1970).[199] They analyzed and tabulated the results in every single case whose diagnosis was confirmed by proper biopsy. They then discussed the putative reasons for success or failure in every instance. This series of monographs was a monumental labor of love undertaken with incredible zeal to save a valuable treatment that they felt was in danger of annihilation.

In all, they reported that Coley and his contemporaries had treated around 2,000 cancer patients. Of these, 896 were microscopically confirmed cases. And of these, 428 (roughly half) were found to be well from 5 to 92 years after treatment. The overall 5-year survival rate was thus 51 percent in operable cases and 46 percent in inoperable ones. This is an extraordinary finding and one that, to this day, has never achieved the attention that it deserves.

The Seven Variables

At first glance, success with Coley's toxins appeared to be haphazard. But upon closer scrutiny, Nauts found, certain well-defined factors made for success or failure in their use. Some of these were under the control of the treating physician, while the patient influenced other factors. (Still others are not understood even at the present time.) The identification of these factors raises the possibility that in the future, Coley's treatment could be made more effective than is indicated in the existing medical records.

Mrs. Nauts' meticulous analysis of her father's cases revealed seven factors that influenced whether or not the patient was likely to be benefited. These were as follows:

1. Stage of disease and/or magnitude of tumor burden
2. Initial immune competence of the patient
3. Potency of various preparations
4. Site of injection, i.e. proximity to tumor, whenever possible
5. Dosage, frequency and duration of injections
6. Timing in relation to surgery and radiation and/or chemotherapy
7. Febrile reactions

Below, I shall examine these seven variables in greater depth.

1. Stage and Burden of Disease

As with any form of cancer treatment, the earlier in the disease process that the patient can be treated, the greater the chance of success. Put another way, patients in the earliest stages of their disease (with smaller and/or non-disseminated tumors) do better than those who already have inoperable or metastatic disease. According to Mrs. Nauts' data, patients who began the Coley treatment when they had primary, operable tumors had a *71 percent* five-year survival rate overall. When metastases were present, however, this was reduced to *28 percent*. Among moribund cases there were *no long-term survivors,* although palliation (reduction in pain, improvement in appetite and weight, better facial color, feelings of well-being, etc.) was often achieved.

We should note that almost all clinical trials of 'new' or nonconventional methods are tested first on advanced, stage IV patients, some of whom could be classified as moribund. Thus, minimal results in such trials may not be indicative of what the treatment could do if administered earlier in the disease process.

2. Immune Competence of Patient

Without adequate numbers of optimally functioning lymphocytes (white blood cells) there is less chance that Coley's toxins will work. The status of the patient's immune system is thus a major deciding issue. Prolonged illness, advancing years, or the cancer itself may weaken the immune system. But nowadays it is more likely that the patient's immune system has been compromised by extensive surgery, radiotherapy and especially chemotherapy. For that reason, there is a possibility that some of today's patients have less likelihood of being cured by this treatment than were patients described in the historical record.

Of course, we must remember that patients respond in individual ways to treatments, and it is difficult to say who is immune compromised and who is not. One sign of compromise, however, is that the patient can no longer react with a fever after the injections of highly antigenic material such as Coley's toxins. This is not uncommonly seen in today's advanced cancer patient. According to Mrs. Nauts, "many of the inoperable or terminal breast cancer patients treated after Coley's death in 1936 received [Coley's toxins] after having had radiation and/or chemotherapy. Although no cures were obtained, marked palliation, regression, decreased pain… and weight gain occurred in many of these patients."

Another confounding factor that could affect the outcome of Coley's treatment is the widespread use of antibiotics in medicine and agriculture since the 1940s. Antibiotics, the "wonder drugs" of the post World War II epoch, are clearly a two-edged sword. Immunity is "a mechanism that has been acquired and perfected through millions of years of evolution" wrote B. A. Meyer and J. D. Benjafield 40 years ago. Yet frequent treatment with antibiotics may interfere with that mechanism. Antibiotics may absolve the body of the need to bring the normal immunological mechanisms into use."[200]

Such criticisms of antibiotics, once associated with the medical fringe, are now more commonly encountered in the mainstream as well. "The widespread and often indiscriminate use of antibiotics" has "created drug-resistant Gram-negative bacilli that readily acquire multiple resistance...." says a typical modern textbook. "Prolonged antibiotic therapy compromises this defense mechanism."[201] It is thus at least feasible that optimal results with Coley's toxins could be diminished by the side effects of long-term use of antibiotics. Again, we should be prepared for the possibility of less dramatic results than were achieved 100 years ago.

One possibility might be for the patient to be prescribed various immune-building substances such as Leukine® (sargramostim, a bone marrow stimulating factor) or any number of milder herbal remedies (such as *Astragalus membranaceus*), as a way of boosting immune response.

3. Potency of Various Preparations

While making Coley's toxins is not a highly technological process, they do have to be prepared with knowledge and care. This has not always been the case. Coley himself was not a microbiologist, and always had to rely on others to make the preparations for him. When those others were knowledgeable and scrupulous technicians, clinical results with Coley's toxins tended to be superior. But when they were not – and sadly this was often the case – then the preparations were either less effective or actually worthless. Coley compensated for some of the weaker preparations with a concomitant increase in the dose administered to patients. But some of the preparations appear in retrospect to have been totally inert. Not surprisingly the long-term remission rates achieved with the various preparations were erratic. This one variable led to a growing sense of confusion, skepticism, and chagrin among Coley's medical colleagues, even some of his erstwhile supporters. This skepticism has been passed down through the decades.

Historically, there have been several different ways of making Coley's toxins. Initially, as we have seen, Coley injected live streptococcal organisms (obtained originally from the laboratory of the famed German "microbe hunter" Robert Koch). This was brutally effective but proved too dangerous to use.

Following this, Coley's colleague, Alexander Lambert, made preparations of toxins for him from *killed* streptococcal cultures. In essence, Lambert first made a beef broth, grew strep in the broth for several days, and then sterilized the product with heat to destroy any living microbes.[202] From 1895 until 1906 Bertram Henry Buxton, MD (1852-1934), Professor of Experimental Pathology at the Loomis Laboratory of Cornell University, made the toxins for Coley. Buxton's idea was to grow the two bacterial species (*Streptococcus pyogenes* and *Serratia marcescens*) together. Buxton retired from Cornell in 1906 and returned to his native England, whereupon Martha Tracy, MD, who had worked with Buxton, continued to make the vaccine until she retired from Cornell in 1923. Between them Buxton and Tracy made what are now considered some of the most effective versions of the toxins.

One thing they learned is that one cannot simply grow these two species together in a haphazard way. *Streptococcus* grows best at human body temperature (37 ° C = 98.6° F), whereas *Serratia* grows best in a somewhat cooler medium. Therefore Buxton grew strep by itself for ten days and only then added *Serratia.* The two were then grown together for another ten days at 25° C (77° F). After 20 days, the preparers killed the microbes by heating them to a temperature of 66° C (151° F).

It was observed that *the lower the temperature used to sterilize the mixture, the stronger the resulting product.* On the other hand, too low a sterilizing temperature could leave some of the microbes alive. The resulting mixture, once injected, could then theoretically infect the patient, with potentially disastrous consequences. Because of this concern, some technicians, in an excess of caution, *overheated the product* in order to make sure that they killed every last microorganism. This may have been smart from a liability point of view, but the resulting overheated product had little anticancer activity. Thus, ironically, some technicians inactivated their own preparation, out of an excess of caution.[203,204]

Martha Tracy, perhaps at Coley's request, also switched to a new formula. There was nothing critically wrong with Buxton's formulation, which had produced many long-term remissions. But around the time of the transition from Buxton to Tracy, Coley became aware that *Serratia marcescens* itself, instead of being just a "prodigious" enhancer of other microbes, might have anticancer activity in its own right. In fact, through experiments on animals, he began to suspect that it was even more powerful than streptococcus.[205]

Martha Tracy's formula was therefore designed to take advantage of Coley's new thinking on *Serratia.* In Buxton's formulas, no one could tell just how much *Serratia* was actually present in the mixture. But *Serratia* turned out to have an extremely variable rate of growth. For this reason, unlike her predecessor, *Tracy grew the two microbes separately* and only mixed them together at the very end of the production process.

Specifically, Tracy took a few cubic centimeters of strep and added these to small flasks of sterilized beef broth. She allowed these to grow in an incubator for three weeks. She then spread some *Serratia* on sterilized agar solution and allowed it to grow at room temperature in daylight (albeit not direct sunlight) for ten days. She then scraped off some of the characteristic red growth of the *Serratia* with glass rods and ground this microbial product with a mortar and pestle into a smooth and thick suspension. She then mixed this with physiological saline (salt) solution, bottled it and sterilized it at 75° C (167° F) for one hour.

Tracy then used what are called 'nitrogen determinations' to figure out how much *Serratia* was present in the resulting product. In this way she was able to achieve "what has never been possible before, a definite standardization of dosage," wrote Coley.[206] All this was admirably precise. However, there was and is also a rough-and-ready way to know whether or not any given ampoule contained much *Serratia.* That is simply to observe its color. Certain strains of *Serratia* are "chromogenic" when grown under the proper conditions and show up as bright red in the vial. The red color is due to a protein, dubbed prodigiosin, which is one of at least a dozen immunostimulatory substances in *Serratia marcescens.*[207,208] The redder the solution, it was said, the greater the concentration of *Serratia*. Reddish products were also found to have greater anti-cancer effects, in line with the emerging realization that *Serratia* itself has anticancer activity.

Although Tracy could measure *Serratia* by measuring the nitrogen content, initially she did not know just how much *Serratia* to include in the final product. She guessed—but unfortunately guessed wrong. "Very severe reactions were obtained by minute doses," Coley reported in 1919, "and in one case, in the hands of another physician, death resulted within a few hours after an injection...into a very vascular tumor in the mediastinal region." This was one of the very rare iatrogenic deaths associated with the use of Coley's toxins.

The amount of *Serratia* was hurriedly reduced by half, and the resulting product was then used safely for decades. There were no more reported fatal reactions. But the lesson was that this essentially simple product had to be produced and administered with great care or very serious consequences could result. While the Tracy product was admirably scientific, it was also somewhat impractical to manufacture under less-than-optimum conditions.

In modern times, after the FDA's ban, when Coley's toxins became a kind of 'underground' cancer treatment, manufacturers reverted to the original Buxton method. The manufacturer (often tutored by a lay enthusiast, Wayne Martin) simply "eyeballed" the pinkness of the final product in order to judge its potency. Before it could be injected into humans, each batch was supposedly tested in animals. Patients then started their treatment with very small doses to test the reaction. The dosage was gradually increased, using the fever/chill reaction as a guide. The proper dose basically depended on the amount necessary to cause the desired fever.

According to Mrs. Nauts, at least 13 different preparations had been used during Coley's active years (1892-1936). *Of these, three were considerably more potent than the rest, Buxton VI, Tracy X and Tracy XI.*[209] The amount of variability of the toxins was startling. For example, in 1902, a patient with recurrent inoperable lymphoma of the pectoral region and axilla reported to Coley that it took 8 minims of Parke Davis formula # IX to achieve the same febrile reaction as 1/4 minim of the Buxton VI. All other things being equal, Parke Davis's formula was 32 times weaker than Buxton's! But despite such clear-cut warning signs, said Mrs. Nauts, Coley "does not seem to have attempted to remedy the situation."[210] This was part of his general failure to standardize the production and use of the toxins in his lifetime. (He was working on a textbook at the time of death.)

The experience of various "underground" clinics (in Mexico, Florida, Guatemala and elsewhere) was that patients had to receive many courses of treatment in order to get on the road to recovery. This led some clinics to adopt the following general guidelines: they started with a minute amount .01 cc (i.e., one-hundredth of a cubic centimeter intramuscularly (IM). They increased this on each successive day by 50 percent and in 13 days they leveled off at 1.00 cc (one cubic centimeter). They then continued the patients on 1.00 cc for a total of four weeks. They then let the patients rest two days and if possible then started them on intravenous (IV) injections at .01 cc increasing this each day by 25 percent for a period of one (1) week. On the following week they return to intramuscular injections, starting at .05 cc and continuing to increase this by 50 percent until they reach a MAXIMUM of 1.00 cc. After two weeks of IMs they return to IVs if possible, starting at the same .01 cc, etc.

Again, this was an empirically derived formula, but was not writ in stone. What if live bacteria were still lurking in the mixture? Could this cause blood infection (septicemia)? In Coley's day, the bacteria were heated to at least 58° C (137° F) and sometimes more. No such problems were encountered. But recent experience has shown that even at 60° C (140° F) some *Serratia* might still be alive in the mixture. Therefore, the formulation made for Wayne Martin at a laboratory in Guatemala was heated to 65° C (149° F) for 2 hours. The resulting product was variable from batch to batch, in terms of efficacy but by all accounts was sterile.

Another controversy concerns the use of preservatives. In Coley's day, preparers added a bit of thymol, a phenol preservative derived from the volatile oil of the herb thyme, to the product. In recent years, some manufacturers have used synthetic thymol for this purpose, while others have used butyl alcohol. MBVax has returned to the use of natural thymol, which they claim is the same as Coley himself used.

Some contemporary mixtures are also filtered. This upsets some Coley purists. Coley himself feared that filtering might eliminate some of the desirable bacterial by-products. However, in Coley's day it did not matter much if a few larger particles were included in the mix, since most injections were given either intramuscularly (IM) or intratumorally (IT). For that reason, both heating and filtering gave a redundant degree of safety. Today, however, when Coley's toxins are sometimes given intravenously, particulates with size greater than one micron can cause serious adverse reactions, even shock.

4. Different Preparations

As I have intimated, there were differences in the methods of preparation and the results that were achieved with different versions of Coley's toxins. Excellent results were often achieved with both the Buxton and the Tracy preparations. But there was also an ineffective commercial preparation of the toxins made at the Lister Institute of Preventive Medicine in England, from about 1895 until 1944. This was designated *Lister VIII* in the literature. (All preparations were given their Roman numerical designations by Mrs. Nauts.) Something was definitely wrong with the Lister Institute's methods of preparation. Even today, it is painful to read how British supporters of Dr. Coley repeatedly wrote to him complaining about the weakness of these British preparations. Injections of this product elicited no fever or chill reactions and concomitantly offered no clinical benefit. To all appearances, they were essentially inert. But Dr. Coley, without our modern means of analysis or communication, was unable to remedy the situation.

Coley crossed the ocean to lecture to the Royal Society of Medicine in 1909 in order to respond to Sir Frederic Eve and other British critics. He was told that "a large number of those present had had unsuccessful results" and that "the closest attention was paid to that part of the address in which Dr Coley explained how much success depended upon a judicious determination of the dosage for a given case." The reasons for failure were argued as follows: "occasionally the physician has heard of the mixed toxins in cases of sarcoma, and if he has read of the method he has probably forgotten the details, and has little faith in its value. Now and then a surgeon is found who takes the time to give the patient the benefit of a trial with the toxins. He begins with some fears and more doubts, and when he sees the patient in one of the severe chills, sometimes with marked cyanosis, very rapid and feeble heart action with a temperature of 103 to 105[° F., ed.], he hesitates to go on or is unable to instill into the patient the courage and confidence so necessary to enable him to continue the treatment to a successful termination."[211]

In the United States, a similar situation (i.e., of an inadequate preparation changing people's minds against the product) existed, especially in regard to the formula now known as *Parke-Davis IX*. The Parke-Davis Company of Detroit (now a division of Pfizer, Inc.) had started production with the Buxton VI strain of strep in 1899.[212] But this strain, being a living microbe, became attenuated over time and lost most of its potency. That was the first problem.

A second was that the company never attempted to standardize *Serratia* in the manner of Dr. Tracy. All in all, it was not a potent or effective formula. Again, American colleagues wrote to Dr. Coley. One possible clue to potency was that some of the bottles were reddish in color, while others were perfectly clear. This may have been a sign that there was little, if any, *Serratia* present in the mix.

According to Prof. Steven Hoption Cann, however, *Serratia* can be quite finicky. Coley said that Dr. Beebe who worked with Tracy, found that *Serratia* "grown in a deep flask, with little surface exposed to light and air, shows little or almost no color, whereas, when the cultures are grown in shallow vessels, with large areas exposed to light and air, the deep red color appears very soon. He [Beebe] has further demonstrated that fact that these red cultures develop much more powerful toxins."

This corresponds to *Serratia* in the environment, Hoption Cann added, which can produce blood red colonies, whereas clinical isolates are clear. With respect to the Parke-Davis formulations, poor quality control and variations in the mass production of the vaccine was probably the reason for their inconsistent product.[213]

Variations in the formula is one of the major reasons that there was a *58 percent five-year survival rate* using Buxton formulas IV, V and VI, a remarkable *67 percent rate using Tracy formulas X and XI*, but, by contrast, a *zero* five-year survival rate with Lister formula VIII (or with the later Sloan-Kettering formula XIV). Everything ultimately depended on the inherent strength of the toxins, although the exact nature of that power was, and remains, a mystery. It may have been its power of prodigiosin or other components to stimulate the production of endotoxin or lipopolysaccharide (as Shear believed) or tumor necrosis factor (TNF). A more recent theory of Kangla Tsung and Jeffrey A. Norton of Stanford University is that Coley's toxins stimulates the production of the cytokine interleukin-12 (IL-12).[214,215] The founders of Coley Pharmaceuticals believe it is a class of DNA compounds called CpG oligonucleotides.[216] One suspects that the immune system, with its network of interrelated factors, is so complex that it might be decades before one can say definitively how Coley's toxins work. But only a dogmatist would say that this precludes the further development of the clinical aspects of the treatment. In any case, we have learned this much: a weak or inactive preparation yields minimal, if any, results, and in fact may make the cancer situation worse.

5. Site of Injection

In chemotherapy, oncologists sometimes deliver toxic drugs as closely as possible to the site of the tumor, even perfusing a limb or organ, such as the liver, with cancer killing chemicals.[217] The same principle applies to Coley therapy. By doing this, an inflammatory reaction is elicited and key elements of the immune system such as macrophages are activated. In fact, in animal studies, the injection of some immunological agents such as BCG (Bacillus Calmette-Guérin) directly into a tumor produces a local inflammatory response, whereas little or no response is seen when the same agent is given systemically.[218]

Initially, Coley himself injected the toxins into any tumor that could be reached by a needle. In this way the inflammatory reactions could immediately be harnessed. However, Coley *suspended* these intra-tumoral injections in 1906, and did not resume them until after he had retired from Memorial Hospital in 1933. The primary reason he had suspended intratumoral delivery was to prove to skeptics that the toxins exerted a *systemic* effect, and did not just act by locally necrosing the tumor in question. This was an important academic point, but should never have been allowed to overshadow the therapeutic needs of the individual patient. By the mid- 1930s, he understood that he had made a tragic mistake in suspending this direction administration of toxins into tumors for nearly three decades—to prove a scientific point—and that by doing so he had also decreased the potential success rate of his treatment.

6. Dosage, Frequency and Duration

A sufficient amount of toxins must be given in order for the patient to mount a proper reaction against his or her tumors. But how much is enough? Generally, as I have stated, the patient should be given the dose that causes a strong fever and/or chill reaction. Basically the particular dose to use at any given time in a given patient has to be established by trial and error. The dose may change, since it can also be expected that a certain tolerance will develop. ***One cautionary note, however, is that too small a dose of Coley's toxins, "given either intramuscularly or subcutaneously, may actually stimulate the target tumor cells rather than inhibit them."***[219]

Ralf Kleef, MD, points out that macrophages have a dual role in tumor initiation and progression, having both pro- and anti-tumor effects.[220,221]

Following the injection, especially when given intravenously, the patient often experiences fever and chills for two hours or more. Some patients, doctors and significant others may have ambivalent feelings about allowing such a reaction and want to promptly reduce it. But through it all one must always keep in mind that this unpleasant fever is the *main point of the treatment.*

Historically, if the patient's temperature averaged 38.9° C (102° F) to 40° C (104° F) or even reached as high as 40.6° C (105° F). If this was accompanied by chills, there was a *60 percent chance* of 5-year survival. This figure does not even take into account other factors that could possibly increase the odds of a response. But when fevers averaged below 38.9° C (102° F), the survival rate in otherwise comparable patients dropped to 28 percent. And if there were few if any fevers, the 5-year survival rate fell further to 20 percent. Thus fever – normally shunned as an undesirable symptom – has a beneficial effect when it occurs as part of Coley's treatment.

Duration of treatment is another especially important factor. In fact, it cannot be emphasized enough (Table 2).

Table 2 Increase in 5-Year Survival Vs. Duration of Administration

Duration of administration	Odds of 5-year survival
1 week	—
2 weeks	14%
8 weeks	42%
26 weeks	80%

To summarize:

• When the toxins were administered for just 1 week, there was no survival advantage to the treatment.
• When they were given for 2 weeks, there was 14 percent 5-year survival.
• When treatment was continued for 2 months, there was 42 percent 5-year survival.
• And when they were delivered for 6 months, which was considered the optimal length of treatment, 80 percent of the patients survived 5 years or more.

Continuing the treatment past the 6-month mark did not increase the survival rate, although a periodic booster shot was considered desirable. **We can conclude from the above that for optimum results the treatment should be given two or three times per week for a minimum of 6 months, especially if fevers of 102° to 105° F can be repeatedly obtained during this time**.

Even many doctors who gave the treatment were unaware of the need to give the treatment aggressively at least *three times per week* initially.

6. Timing in Relation to Surgery

Understandably, because it never was accepted as a conventional treatment, Coley's toxins were generally given in cases that were very far advanced, even moribund. In such cases, although it still could possibly cause long-term survival, more frequently it simply palliated the condition. It should be emphasized that Coley's toxins in no way contradicted the judicious use of surgery. In fact, it was probably most appropriate when used as an adjunct before and after conservative surgery.

Mrs. Nauts pointed out that it is "particularly needed in all postmenopausal women to augment their response to adjuvant chemotherapy." She felt it should be given during the time when recurrences or metastases were most likely to develop. It could be used to reduce the tumor burden and to protect against the harmful effects of chemotherapy and radiation "while potentiating the response of the tumor to these modalities."[222]

7. Fevers and Chills

Fever and chills are the hallmarks of Coley's treatment. They are certainly unpleasant, but are they harmful or should aggressive measures be taken to reduce them? Opinions differ. According to the *Merck Manual*, "whether to treat an elevated temperature that occurs with an infectious disease is the subject of an ongoing debate. Experimental evidence suggests that host defense mechanisms are enhanced by an elevated temperature; thus, fever is potentially beneficial and should not be routinely suppressed."[223]

In cancer patients, the development of fever can be a serious complication. This is especially so when the patient is suffering from immune suppression following chemotherapy. I am not commenting here on the wisdom of antibiotic therapy in such cases. I am just trying to point out that there is a strong philosophical bias against not just allowing fevers or utilizing them as a therapeutic procedure.

It bears repeating that patients who had fevers which averaged between 102° and 105° F (39–40.5° C) had a significantly higher percentage of complete and permanent regressions. This was especially true in inoperable or metastasized cases. In addition, the responses were better when injections were given in or near the tumor; intravenously, or in larger intramuscular doses. The subcutaneous route was historically the least effective. The largest number of "spontaneous remissions" also followed infections in which there was a high fever. But fevers are not easy to endure. If the fever goes on beyond two hours it can be brought down within fifteen minutes or so by acetaminophen suppositories or other anti-pyretics.

It is well known that infections, whether naturally occurring or (as in the case of Coley's toxins) simulated, can trigger the body to release compounds known as endogenous pyrogens, which are substances that cause fever. The main one of these is interleukin-1, which is released by a specialized macrophage of the liver sinusoids known as the Kupfer cell. But other pro-inflammatory cytokines, such as interleukin-6 (IL-6) and tumor necrosis factor-alpha (TNF-a) are also involved. The presence of fever in response to Coley's toxins is a reliable sign that these three pro-inflammatory cytokines (and others as well) are being released.

An increased presence of these cytokines can in turn stimulate various important immune function cells, including the activation of monocytes, macrophages, keratinocytes, and endothelial, B, epithelial and glial cells. These in turn influence the thermogenic (heat-generating) center of the hypothalamus, which is sometimes called the "master gland" of the body.

Macrophages not only produce tumor necrosis factor (TNF) and other cytokines but also directly kill cancer cells. However, they must first be activated by other factors. The most effective way of activating them is to bring them into contact with "a variety of microorganisms and their structural components."[224]

Various involved and expensive ways have been devised of doing this. For a while there was a vogue of administering very high doses of recombinant cytokines (especially interferon-a and interleukin-2) to cancer patients. According to Hoption Cann, these so-called 'immune stimulating' cytokines can also cause neutropenia, and ironically, increase a patient's risk of infection. This, however, is not an adverse effect of Coley's toxins.

Coley's toxins, on the other hand, remain a relatively easy and inexpensive way of stimulating the body to produce its own supply of cytokines. It is more natural than recombinant products because it simulates something that happens on a regular basis in our species, i.e., bacterial infections.

Toxicity

One of the historically unfortunate things about Coley's treatment was the choice of the name "toxins" to apply to it. Many people over the years have attempted to devise a less frightening sound appellation, such as "mixed bacterial vaccine." They have never been entirely successful. The name "Coley's toxins" is memorable and, for better or for worse, has stuck.

But how 'toxic' are Coley's toxins? The answer depends on what one means by "toxic." The goal of this treatment is to stimulate an immune reaction such as one would normally experience if assailed by an intense bacterial infection. Fever, chills, sweats, or even headaches can be very unpleasant, and in that sense one must acknowledge that this certainly is a treatment with seriously unpleasant side effects.

Plus, we must remember that Coley himself was aiming for the maximum therapeutic effect and to get this he sought out the most virulent strains of *Streptococcus pyogenes* and *Serratia marcescens* that he could find. As he once said, "I obtained cultures from fatal cases of erysipelas in order to get the highest degree of virulence." In fact, one strain of strep that he particularly cherished came from the Collis P. Huntington Fund for Cancer Research of Memorial Hospital, where it was isolated from a fatal case of septicemia.

However, that said, we must note that the word "toxic" has changed its meaning somewhat in the intervening decades. Chemotherapy has added a new dimension of toxicity to the medical textbooks. Much chemotherapy has a tendency to destroy normal cells, especially those of the immune system. By contrast, to our knowledge, Coley's toxins leave the various organ systems and tissues of the body in an equal, if not improved, state at the end of the treatment than they were at the beginning.

"The available evidence suggests that Coley's toxins are without harmful or dangerous effects to patients or animals suffering from various types of neoplasms, provided these toxins are administered properly as to dosage, site and the usual aseptic precautions." Helen Coley Nauts did emphasize the following limitations on the treatment, however, which should be understood by all those considering this route:

> "They should not be given to patients with severe hepatic [liver] insufficiency due to metastatic disease or other pathology, nor to patients who have had severe heart conditions, nor to patients who are almost moribund, because such patients do not respond." However, any questions on the potential danger of the treatment must be referred to a competent medical doctor, especially one with first-hand knowledge of this type of treatment.

Ralf Kleef, MD, has commented: "I only partially agree with Helen Coley Nauts on this point: I believe that liver metastases are no contraindication as long as the liver function (e.g. level of cholinesterase within normal range) is sufficient."[225]

Clinical Trial Results

Many people, when they first learn of Coley's toxins, think that surely 100 years of intense controversy would have given rise to definitive clinical trials...and credible answers. Clinical trials have been performed, but they have not settled the issue to the satisfaction of most physicians.

At a fateful meeting at Memorial Hospital on May 25, 1934 Coley for the last time presented many cases of cured bone cancer. He then contrasted these cases with what he perceived as the failure of the radiation therapy of the day in similar cases. This put him into direct conflict with Ewing, who was regarded as a great expert on these tumors (a particular type of which even bore his name).

Ewing was as devoted to the promotion of radiation treatment as he was hostile to Coley's unorthodox treatment. He also had other strong opinions, some of which seem strange today. For instance, he was adamantly opposed to needle biopsies, which he felt spread cancer. This definitely put him in the minority compared to Coley and others who favored preoperative biopsies. In fact, Ewing went so far as to suggest, at this 1934 meeting, that – without biopsy proof of cancer – "every case of persistent unexplained pain in a bone should be regarded provisionally as sarcoma and treated by radiation."[226]

Coley spoke to this statement in his last paper: "I cannot subscribe to Ewing's statement...We have no evidence at present to prove that radiation offers any reasonable hope of a cure, no matter how early such treatment is given." He added: "we are no longer justified in advocating primary irradiation in the early stages of osteogenic sarcoma...certainly not without fully informing the patient and his family that surgical treatment alone, or combined with Coley's toxins, offers from 15 to 30 percent chance of five-year cure, whereas irradiation alone had failed to cure more than one or two cases in a very large group in which this method was employed."[227]

The prospect of every bone pain sufferer being treated aggressively with x-rays might have seemed attractive to a hospital administrator at the height of the Great Depression (when Memorial's radiation department was the most profitable part of the hospital). But, even in those days, it must have seemed like an outrageous proposal. After all, a lot of people have bone pain, for a variety of reasons, but few of them have bone cancer. Luckily, Ewing's suggestion never was adopted, since most doctors would not treat patients for cancer without biopsy confirmation of cancer or some other dire disease.

But the most surprising outcome of this fateful meeting was the "conversion" of a famous Boston surgeon Ernest A. Codman, MD, (1869-1940) from pesky critic to ardent believer in Coley's work. Codman, who was a pioneer of outcome-based evaluation of doctors, had previously driven Coley almost to distraction with persistent criticism. (Helen Coley Nauts attributed her father's ulcers in part to Codman's acerbic letters.) But at this 1934 meeting he suddenly acknowledged the "not infrequent miracles" achieved with the toxins. Codman said: "Some great clinic should try out Coley's toxins during the next decade."[228] Here was the key point – a proposed large-scale clinical trial. But for various reasons, this was not done. "It was an opportunity lost, probably forever," wrote Hall.

Such a clinical trial could have been done at Memorial, Massachusetts General Hospital (with which Codman had once been affiliated) or the Mayo Clinic, which used Coley's toxins almost as extensively as Coley himself did. But no one picked up this challenge.

Coley himself suggested a few reasons that the toxins failed to gain traction. We must remember that at that time the concept of a medical oncologist did not exist; virtually all cancer care was in the hands of surgeons. These, basically for economic reasons, wanted a quick turnover of patients, not prolonged stays of weeks and months while the toxins were patiently administered and calibrated to each individual. Patients and their family doctors resisted the extra cost of the treatment (especially during the Great Depression); and surgeons generally had no experience in the use of the toxins and were unwilling to spend time learning about them.

Although Coley published 143 papers or monographs on his method between 1893 and 1936, these did not give sufficient detail, said his daughter, "on how to administer the toxins, i.e. site, dosage, frequency and duration, and the optimum febrile reaction to aim for."[229] He was working on a textbook on the toxins at the time of his death. This could have brought together in one place the principles and practice of his mixed bacterial vaccine therapy. When his son, Bradley, published a textbook on bone cancers in the late 1940s it contained just one slim chapter on the toxins, assigning them a minor role compared to standard surgery and radiation.

Equally damaging was that Dr. Coley did not exercise rigorous quality control over the various products that bore his name, especially those produced by Parke-Davis from 1898 until they abruptly halted production by 1953.[230] Coley himself had no bacteriological training, and so he relied on others to make the various preparations. But he and his collaborators never standardized the dose, preferable route of administration, or most likely indications for the treatment. These issues, inadequately explained in his day, have dogged the use of these biological products since then.

More fundamentally, some of the implications of Coley's treatment went counter to deep-seated, nearly unconscious presupposition about good medicine. Increasingly, the focus of cancer treatment was on ablating or destroying tumors through strenuous human effort—exemplified by "heroic" surgery. (The ultimate manifestation of this was the hemicorporectomy, in which surgeons removed the lower half of the human body in a valiant attempt to save the upper half.[231])

Coley's emphasis, by contrast, was the intense stimulation of the body's own healing powers (the so-called 'vis mediatrix naturae'). This is consonant with the essential philosophical concept of naturopathy. During Coley's lifetime, some non-conventional practitioners, such the San Bernardino naturopath, Emerson B. Hartman, offered an "antitoxin that has cured the worst cancers known." According to *Time* magazine, "the specious idea behind it skulks in the shadow of the very real cures of certain kinds of bone cancer which Dr. Coley has been able to make with a toxin."[232] Coley was on dangerous turf, however, when his own treatment could be confused with those of unlicensed naturopaths.

Coley was also exploring an area of science that was in its infancy, cancer immunology. Until the pioneering work of Prof. Jacques Miller of Melbourne, little was known about the role of the thymus in immunity or the different classes of lymphocytes in the blood.[233] There was a general assumption – well into the 1960s – that lymphocytes were of secondary importance in immunity. We have come a long way, yet, even today, the precise mechanism by which Coley's toxins work remains unknown. This weighs heavily on the mind of many who might otherwise be more sympathetic to the concept of cancer immunotherapy.

All Too Human

Finally, there was the "all too human" aspect of science. The successes scored by Coley's toxins represented a competitive challenge to other would-be adjunctive treatments, particularly radiotherapy.[234] James Ewing, who was a commanding advocate of radiotherapy, blocked Coley and his toxins at many points. His opposition – followed by that of Cornelius P. "Dusty" Rhoads, and other Memorial Sloan-Kettering leaders in the postwar period– obstructed the path to a fair and comprehensive clinical trial, that could have definitively established a role for the toxins in the postwar era.

In 1962-63, the Food and Drug Administration (FDA) was given enhanced powers under the Kefauver-Harris amendments.[235] Soon thereafter the FDA refused to 'grandfather' Coley's product, as it had done with aspirin and other traditional products, but insisted on extensive (and expensive) clinical trials in order to approve it as a new drug. These were beyond the means of Mrs. Nauts and her few supporters in the Cancer Research Institute. So, from that point, Coley's toxins were effectively banned in the US outside the context of clinical trials. They still cannot be manufactured for sale in the US. Although now made in Canada for use in a dozen countries, they cannot be shipped across the US border, where they would be regarded as an unapproved, and therefore, illegal drug and seized.

Nonetheless, research has never stopped on the toxins, even in the US. The sterling track record, attested to by Mrs. Nauts and Dr. Old, continues to attract immunotherapists. In every decade since Coley's death there have been attempts made to prepare the ground for approval, although none of these ever successfully met FDA's stringent requirements.

Clinical Trials

It is sometimes assumed that there has been no clinical trial performed on mixed bacterial vaccine since Coley's day. This is a misconception. There have been trials performed in every decade since the 1950s and all of them have been more or less positive. But it is also true that, for various reasons, none of these trials has been large or convincing enough to pass muster with the Food and Drug Administration (FDA).

Let us review the trials that have been conducted since the 1950s. The monograph, *Mechanism and Studies*, summarizes the various Coley trials. This excellent treatise was primarily written by Prof. Stephen A. Hoption Cann, School of Population and Public Health, University of British Columbia and is available at the Web site of MBVax Bioscience (see below).

The Johnson Trial[236]

From 1954-1957, a research group at New York University-Bellevue Medical Center, New York City, under the direction of Barbara Johnson, MD, received a grant from the National Cancer Institute (NCI) to conduct a clinical trial of Coley's toxins. This resulted in two papers, which were published consecutively in *Cancer Chemotherapy Reports*.[237]

Thirty-four (34) patients with inoperable metastatic cancers were treated with Coley's toxins. These were administered intravenously 5 days per week, unless tumors were easily accessible in which case intratumoral injections were administered every 4th or 5th injection day. (The initial dose was 0.05 ml of Coley's toxins in a 1:20 dilution.) Doses were increased incrementally to produce a fever between 38.8° C (101.8° F) and 40.0° C (104.0° F). Coley's toxins were initially administered for 3 weeks, and then continued for subsequent cycles 3 times per week, unless there was disease progression.

An additional 37 patients, serving as a control group, received a typhoid vaccine, but were otherwise treated in the same manner. Patients were randomized to one of the two groups and given identical nursing care and supportive therapy. No restrictions were placed on either food or fluid intake during therapy.

Only one of the typhoid treated patients showed any improvement, which was of short duration and questionable significance. In the group treated with Coley's toxins, 16 of 34 patients showed some improvement. Of these 16, nine showed objective improvement. The others had subjective changes, such as decreased pain or reduced need for narcotics. One patient showed a significant increase in weight (15 lbs) and two had decreased bleeding. All patients had severe chills and fevers, as expected. Major side effects were occasional dyspnea and cyanosis (bluish discoloration of the skin) during their chills, occasional pain in the area of the tumor, as well as occasional nausea and vomiting. Nausea and vomiting were actually more common in the typhoid treated group.

There was no depression of marrow activity, white cells or platelets. Occasionally there was a drop in hemoglobin or hematocrit (i.e., the ratio of the volume occupied by packed red blood cells to the volume of the whole blood), but only in those with bleeding to explain blood loss.

The Second Johnson Study[238]

This study followed the previous randomized Johnston trial. It was undertaken to identify the particular types of tumors that might respond best to Coley's toxins. All 93 patients in this study had progressing cancers, had been off all other therapy for at least two and one half months and conventional modes of therapy were no longer effective in these patients.

Only patients receiving more than 10 injections of Coley's toxins were included in the analysis. The product was administered intravenously 5 days per week, unless tumors were easily accessible in which case intra-tumoral injections were administered every 4th or 5th injection day. The initial dose was 0.05 ml of the product in a 1:20 dilution. Doses were increased incrementally to produce a fever between 38.8° C (101.8° F) and 40.0° C (104.0° F). As in the randomized trial, the toxins were administered for 3 weeks, and then continued for subsequent cycles 3 times per week unless there was disease progression. No restrictions were placed on food or fluid intake during therapy.

Objective improvement occurred in 30 out of 93 patients (32.3 percent), which consisted of decrease in size of lesions, involved lymph nodes or liver disease. No improvement was seen in 43 out of 93 (46.2 percent) of the patients. In the remaining 50 (53.8 percent) there was subjective improvement, primarily decreased pain. Three showed a gain in weight in addition to reduced pain. Five patients had long-term survival benefit.

The Chandler Trial[239]

James J. Chandler, MD, and colleagues in the Department of Surgery, UMDNJ-Robert Wood Johnson School of Medicine, Piscataway, NJ, studied Coley's toxins in 21 patients with various tumors. The toxins were injected intravenously and, when possible, intratumorally as well. The starting dose and dosing increments were not given in the article, but were presumably similar to those in the Johnson trial (above). The dosage was increased gradually in order to produce a fever in the range of 38.3° C (101.0° F) to 40.3° C (104.5° F).

Patients received injections 5 days a week for three weeks followed by biweekly to weekly injections. Three patient groups were treated. The first group included six patients with operable cancers including melanomas and sarcomas, which also received surgery and, in some patients, chemotherapy or radiotherapy.

One patient with melanoma of the left temporal region was treated with surgery + Coley's toxins (45 injections) and was clinically free of disease at the time of reporting two years later.

One other patient with melanoma was well following discharge receiving a limited number of injections (15 injections of Coley's toxins) and surgery. This patient had a surgically treated recurrence one year later, but was well at the time of reporting. The other 4 patients died of their disease with no objective benefit from the treatment.

In the second group, Coley's toxins were used as palliative treatment in seven patients following the failure of conventional treatment and advancement of disease. All patients but one received a very limited number of injections and died of their disease. One patient with melanoma who received 52 injections over a period of 14 months was in good health at the time of reporting, but had multiple nodules present. In a third group, eight patients with advanced disease and tumors that were considered radio-resistant were given Coley's toxins followed by palliative radiation to localized symptomatic areas. Six of eight patients were considered to have definitely benefited. All but one patient subsequently succumbed to their disease.

One patient with a fibrosarcoma who had a dramatic reduction and disappearance of multiple tumor masses was alive at the time of reporting, with some residual evidence of malignant disease. The reported side effects of Coley's toxins included fever, chills, fatigue, and pain and increased temperature at the tumor site following injection. This occurred commonly in non-injected lesions and it was noted that Coley's toxins often aided in revealing lesions not previously apparent. The symptomatic effects of Coley's toxins were reported to be short-lived and resolved in several hours as the temperature fell. In one case there was an episode of moderate hypotension that was considered dose related. The effects of Coley's toxins were reported to be mild in comparison to the side effects observed following chemotherapy.

The Havas Trial[240]

Starting in the early 1960s, Prof. H. Frances Havas of Temple University conducted pre-clinical experiments with Coley's toxins, under the name mixed bacterial vaccine (MBV).[241,242,243,244] These experiments culminated in a clinical trial, published in 1993.[245] The purpose of the trial was to assess the impact of Coley's toxins on both malignant growths and on the typical immunological parameters studied at that time: interleukin-2 (IL-2), interferon (IFN) and tumor necrosis factor (TNF).

Eleven patients with various kinds of cancer, including one with Kaposi's sarcoma and HIV/AIDS, who had either failed or refused conventional therapy or for whom no conventional therapy was available were studied. Levels of TNF were undetectable for all patients at the start of the study. Coley's toxins were given (as a single agent for all except the AIDS patient) initially intravenously and subsequently by intramuscular injection twice per week. The treatment was administered in a range of 51 to 457 days. One ml of the toxins was diluted into 100 ml normal saline for the intravenous injections, while a 1:10 dilution was used for intramuscular injections.

No patient achieved a complete remission (defined as complete disappearance of disease for a month or more). One patient had a partial response (defined as a decrease of at least 50% in the measurable areas of the disease with no appearance of new lesions or progression of other lesions). Four patients had stable disease, one patient had a minor response, while the remaining five patients had progressive disease.

There was mild fever and chills following intravenous administration of Coley's toxins but there were no adverse reactions to the injections.

The interferon levels rose for three of the eight patients tested but dropped again for two of these patients. TNF levels rose to a detectable level for one patient and to a high level for another patient; however, this latter patient also had AIDS and was also taking AZT. IL-2 levels rose and surpassed the levels of two normal control patients for four patients but dropped again for three of these patients. (No statistical analysis of the immune data was performed due to the diverse patient population and different number of treatments and assays.) The performance status as measured on the Karnofsky scale improved from 60 percent to 80 percent for one patient and bone pain completely disappeared for another patient. Coley's toxins were deemed to be "a well-tolerated biological response modifier with modest activity in advanced human tumors."[246]

The Axelrod Series[247]

This is the case series produced by Rita Axelrod, MD, a well-known research physician in Philadelphia, who was at the time a member of Frances Havas' group at Temple University. Twelve patients were evaluated following treatment with one or more cycles of Coley's toxins. They had colorectal cancer (1 case), pancreatic cancer (1), chronic lymphatic leukemia (1), hepatoma (2), and sarcoma (3) and were treated with Coley's toxins alone (0.1 ml in 100 ml normal saline intravenously) twice per week for 3-16 weeks.

Four patients with advanced non-small cell lung cancer were also treated with Coley's toxins in combination with chemotherapy.[xvi] This cycle was repeated every 28 days. Plasma interferon levels, interleukin-2 production by peripheral lymphocytes, and lymphocyte subpopulations were also monitored.

Survival after the start of treatment in this refractory cancer group ranged from 34-650 days (the latter patient being still alive at the time of reporting). Mean survival was 197 days. In the group with lung cancer, survival in the Coley's toxins group ranged from 105-306 days (the latter alive at the time of reporting) with a mean of 279 days, while in the control group, survival was 16 and 19 days. Adverse effects included, not surprisingly, chills (68%) and fever (23%), which were almost certainly desirable under the circumstances. There was also an increased respiratory rate (33%), blood pressure increase (25%) or decrease (5%), nausea (11%), and in less than 5%, wheezing, cyanosis, vomiting, and headache. Interferon levels and interleukin-2 production showed increased or sustained values in general. In some patients, B-cells and helper T-cell populations also increased, whereas T-suppressor cell numbers declined. In this pilot study, then, cancer patients treated with Coley's toxins showed objective evidence of immune stimulation with acceptable levels of toxicity.

Kempin Trial[248]

In 1983, Sanford J. Kempin, MD, and colleagues in the Department of Hematology and Oncology, Saint Vincent's Hospital, Manhattan conducted a clinical trial with Coley's toxins. Although an abstract was published in the *Proceedings of the American Society of Clinical Oncology,* but unfortunately an article on the experiment was never produced for any PubMed-listed, peer-reviewed journal.[249] The purpose of the study was to evaluate the effect of Coley's toxins (called mixed bacterial vaccine, or MBV) on the response and survival of patients with advanced nodular non-Hodgkin's lymphoma. Fifty-six patients with this diagnosis (in stages II-IV) were treated with the so-called NHL-4 protocol.

Coley's toxins (MBV) were administered in randomized fashion for just 1 week prior to chemotherapy. There were 26 patients in the therapy group and 30 in the control. The overall complete (CR) and partial response rates were 62% and 32%, respectively (see Table 3). The length of survival of patients treated with MBV was also significantly better than for non-MBV patients (p= 0.029); however the median follow-up was only 34.3 months.

Table 3 Nodular Lymphoma Trial Results

[xvi] Low-dose cyclophosphamide, day 1; cisplatin, day 15; and CF, 0.1 ml intravenously, days 5, 7, and 9. Two control subjects received cyclophosphamide and cisplatin alone.

Method	*# of patients*	*CR + PR rates*
Chemo + MBV	26	62%
Chemo alone	30	32%

Data: Kempin (1983)

There are uncertainties about this trial, because only the abstract survives, while full results were never published in a journal. We do not know what longer follow-up might have revealed.

Chinese Trials

In 1981 Dr. Z. Guo, a pediatric oncology surgeon at Beijing Children's Hospital, became interested in using Coleys toxins for his patients. Mrs. Nauts reported that he was impressed by the case of a patient with a very extensive sarcoma of the thigh that had completely recovered following a severe staphylococcus infection. He thereupon read one of Mrs. Nauts's monographs, which was in the Medical School Library. Mrs. Nauts then went to China in October 1983 and brought Dr. Guo to New York that fall to visit pediatric oncologists at Memorial Sloan-Kettering Cancer Center in New York. She also provided him with the toxins prepared by Frances Havas in Philadelphia.

Mrs. Nauts reported that in 7 and 1/2 years, Dr. Guo had treated 49 cases in children. Of the inoperable cases receiving 10 or more injections 8 were successful, traced up to 8 years after onset. There were 16 inoperable failures, six of whom received just 7 to 10 injections—too brief a period, she commented. Of the 22 operable cases, only two have died, one of which was recurrent when toxins were begun; 12 have remained free from disease up to six years after onset; seven more recent cases remain well and one equivocal case is probably cured but awaits a "second look."

In the late 1980s, Zhou-You Tang, MD, chairman of the Liver Cancer Institute of the Zhong Shan Hospital, Shanghai Medical University, Peoples Republic of China, conducted a small clinical trial to investigate the effects of mixed bacterial vaccine in patients with hepatocellular carcinoma, the most common form of liver cancer.[250] Eighty-six patients with pathologically proven HCC were entered into the trial from March 1985 to June 1988. Y.T. Zhao and colleagues published their preliminary results on the use of mixed bacterial vaccine in the treatment of liver cancer.[251]

This randomized controlled trials (RCT) had two separate cohorts. In the first, 38 patients received palliative resection of their tumors, but half (19) also received Coley's toxins, while half did not. In the second cohort, 48 patients with unresectable (inoperable) cancer all received hepatic artery plus other intraarterial infusion of the drug cisplatin. Twenty-five of these then received MBV (i.e., Coley's toxins), while 23 served as controls (see Tables 4 and 5).

Table 4

Survival in Cohort #1, Palliative Resection plus or minus MBV

	# of patients	One year survival	Two year survival
MBV (Coley)	19	75	45
Controls	19	58	39
Significance	—	P = 0.19	P= 0.23

Data: Tang, et al. 1991.

Table 5
Survival in Cohort #2, Unresectable HCC, w/ or w/o MBV

	# of patients	1-yr survival	2-yr survival	3-yr survival
MBV Group	25	59	41	41
Control	23	39	25	20
Significance	—	P = 0.07	P = 0.09	P = 0.07

Data: Tang, 1991.

Three-year survival with unresectable liver cancer was more than twice as common among those who got the Coley's toxins (MBV) as among those who got conventional therapy alone. In every case, the data trended towards being positive, however the small sample size precluded formal statistical significance. Nonetheless, nothing further appears to have happened as a result of this trial.

The Kölmel Pilot Study

In the 1980s and 1990s, Klaus Kölmel, MD, a melanoma specialist at the University of Göttingen, Germany, conducted a pilot study with a "pyrogenic bacterial lysate," i.e., Coley's toxins.[252] Fifteen patients with metastasizing melanoma (8 female and 7 male) were injected with Coley's toxins. Depending on their tolerance to the first injection, the intravenous application continued weekly up to a total of 12 injections. In case of progression, treatment was discontinued after the 6th injection. In case of remission, the treatment was continued longer than 12 injections, until a complete remission was observed. In 3 of the 15 cases, Coley's toxins resulted in a total and long lasting remission. (The three cases all had skin metastases). In another case with inguinal lymph node metastases, a 5-month period of stability was achieved. The remaining 11 patients showed further progression of the disease. Side effects included fever, nausea, headache, back pain and occasionally herpes of the lips. TNF-α levels were elevated with temperature. Marked increase in monocytes was noted 24-48 hours after injection. None of the side effects (fever, chills, diarrhea, etc.) were long lasting, nor did they cause an interruption of treatment.

The Bajenov Trial[253]

In March 2006, Leonid G. Bajenov, MD, PhD, and colleagues reported on the analgesic effects of Coley's toxins at the 2nd World Congress on Regional Anaesthesia and Pain Therapy in Rio de Janeiro. Dr. Bajenov is a professor at the Vakhidov Research Center of Surgery, Tashkent, Republic of Uzbekistan.

This study was carried out on 16 patients who were experiencing extreme pain from cancer of the stomach, liver and pancreas. For 8 of these patients, standard analgesic drugs provided only moderate pain relief. Patients were treated primarily with intramuscular or intratumoral injections. By the end of the first course, 5 (31.3%) discontinued all analgesic medications, while the remainder obtained varying degrees of analgesic dose reduction. In general, these patients were observed to have an improved appetite and sense of well-being, improved biochemical and clinical parameters, and at least partial tumor regressions were observed in 7 (43.8%) of the patients.[254]

The purpose of the Bajenov study was to examine the effect of Coley's toxins on cancer pain and tumor regression. It was used in patients with stage 3 or 4 cancer, primarily cancer of the stomach, liver and pancreas. The study focused on patients with severe cancer-associated pain. For 8 out of 16 of these patients analgesic drugs provided only moderate pain relief. Patients were treated primarily with intramuscular injections, and with intratumoral injections whenever possible.[xvii]

By the end of the first course of treatment, 5 out of 16 (31.3 perent) had discontinued all of their analgesic medications, and varying degrees of analgesic dose reductions were obtained by the remainder of the patients. In general, patients also had an improved appetite and sense of well-being, improved biochemical and clinical parameters, with partial tumor regression in 7 of 16 (43.8%) patients.

Planned Trials

There are, to my knowledge, two trials of Coley's toxins in the planning stage.

Nordwest Trial

This is a phase I trial of mixed bacterial vaccine at the Krankenhaus Nordwest in Frankfurt, Germany, sponsored by the Ludwig Institute of Cancer Research (a well endowed foundation, of which Lloyd Old is chairman of the board).[255] The key researcher is Prof. Elke Jaeger, MD.

[xvii] Concentrations of the toxins were increased from 0.01 ml up to 3-5 ml within 21-28 days, following a 1-2 weeks rest period, the course of treatment was repeated.

Danish Trial

A Danish combined phase 1 and 2 clinical trial of MBVax Coley Fluid in sarcoma, melanoma and bladder cancer is scheduled to begin in the first half of 2009.[256]

Systematic Reviews and Discussions

In addition to the abovementioned clinical trials, in recent years there have been half a dozen or so systematic literature reviews of Coley's results, using modern methods of data analysis. Although couched in cautious terms, all of these have come to generally positive conclusions about the potential of Coley's toxins to once again become a viable treatment option. I view these authors, who are located all around the world, as heroic, for it is not easy to spend time and energy on a topic that some scientific colleagues regard as eccentric, frivolous, or worse, and that carries little chance of material success.

Here are samples from their excellent papers' conclusions, with my comments:

- *1994:* "Using a bacterial vaccine to treat primarily inoperable sarcoma. Coley accomplished a cure rate of better than 10 percent." Bernadette Wiemann and Charlie O. Starnes, Amgen, Thousand Oaks, Calif.[257]

Charlie Starnes was one of the first scientists (outside Helen Coley Nauts's circle) to take Coley's work seriously. The appearance of this article was noteworthy at the time because it heralded a renewed interest on the part of scientists in Coley's ideas.

- *1999:* "[P]atients treated with surgery and Coley toxins between 1890 and 1960 experienced survival rates comparable to those of patients diagnosed in 1983 and treated with nonradiotherapeutic conventional approaches." Mary Ann Richardson, et al, M.D. Anderson Cancer Center, Houston.[258]

Mary Ann Richardson, PhD, was a pioneer in bringing the first investigations of CAM to M.D. Anderson Cancer Center. Richardson then moved on to the National Center for Complementary and Alternative Medicine (NCCAM) and the National Foundation for Alternative Medicine (NFAM).

- *2002*: "The best response by far was achieved in patients with inoperable soft tissue sarcomas; long-term (more than 5 years) disease-free survival was achieved in approximately 50% of these patients." —Jacob Bickels et al, Tel Aviv Sourasky Medical Center (2002).[259]

Jacob Bickels is an attending orthopedic oncology surgeon at the National Unit of Orthopedic Oncology, Tel Aviv Sourasky Medical Center, Tel Aviv, Israel. He completed an orthopedic oncology fellowship at Washington Cancer Institute. He thus is among the first modern orthopedic oncologists to take an interest in the work of one of the founders of his field. His orthopedic coauthors were Yehuda Kollender, MD, Ofer Merinsky, MD and Isaac Meller MD.

- *2001*: "Pyrogenic [fever-inducing, ed.] substances…have successfully been administered in palliative and curative treatment protocols for metastatic cancer." —Ralf Kleef, MD, et al, National Institutes of Health.[260]

From 1994 to 1996, Ralf Kleef, MD, was a Cancer Research Institute postdoctoral fellow at Sloan-Kettering Institute 1994 in the laboratory of Dana Bovbjerg. He also worked closely with Lloyd Old. After this, he went to the National Institutes of Health (NIH) where he produced this paper on Coley's toxins with Wayne Jonas, MD, then director of the Office of Alternative Medicine (now the National Center for Complementary and Alternative Medicine, NCCAM). Dr. Kleef now directs the Institute for Hyperthermia and Immune Therapy (*Institüt für Warme und Immuntherapie*) in Vienna, Austria, using modern-day immunological methods to fight cancer.

- *2001:* "Fever induction under medical guidance may be considered as part of a therapy regimen for cancers of mesodermal origin." Uwe Hobohm, University of Giessen, Germany.

Uwe Hobohm is a professor in Giessen, who has written outstanding articles bringing Coley's work into the modern world. He thus continues the 150-year German interest in the use of erysipelas and erysipelas-based products in cancer. His classic 2001 paper on "Fever and Cancer in Perspective" was the first to discuss spontaneous regressions, post-Coley studies and the role of fever together under one umbrella.[261] He was also the first to generate an immunological hypothesis for Coley's toxins, namely, the unspecific activation of dendritic cells, which he still believes to be valid.[262] This paper was then combined with the epidemiological evidence, again for the first time, in 2005 in "Fever Therapy Revisited." [263] He then put all this evidence into molecular biological perspective in a 2008 paper, "Pathogen Associated Molecular Pattern" (PAMP) in Cancer Immunotherapy."[264]

At the same time, Hobohm has written some of the best articles for non-specialist readers on the relevance of Coley's work.

- "[T]he modern science of immunology has shown that Coley's principles were correct and that some cancers are sensitive to an enhanced immune system. Because research is very active in this field, William B. Coley, a bone sarcoma surgeon, deserves the title 'Father of Immunotherapy.'" —Edward F. McCarthy, MD, Johns Hopkins Medical Institution, Baltimore (2006)[265]

McCarthy's primary appointment at Johns Hopkins is in bone pathology, his secondary appointment in orthopedic surgery. He has a bone biopsy histomorphology laboratory that is available for diagnostic problems and research. He is interested in various bone diseases, including primary and secondary bone tumors. He is studying certain bone tumors, particularly cartilage tumors, with respect to rate of turnover as depicted in immunochemical techniques. He is thus the second orthopedist to express an interest in the work of Coley, a foundational figure in their field.

Thus, an outstanding group of researchers are determined to see Coley's treatment reevaluated and resuscitated in the 21st century. Yet this is decidedly not the mainstream opinion. In fact, a search of the past ten years of the American Society of Clinical Oncology (ASCO) annual meetings is quite sobering: there are *no* abstracts that mention the words 'Coley's toxins' or 'mixed bacterial vaccine.' This is startling, when one realizes that there are over 20,000 presentations made at each meeting.[266]

The Web site of the American Association for Cancer Research (AACR) is not much better. Coley's photograph is dutifully shown as one of a dozen founders of the organization, and there is a passing reference to what they call his "primitive" vaccine, but there is not a word about the actual nature of his treatment, much less the clinical results he achieved.

For all intents and purposes, then, Coley's toxins have fallen off the map of conventional oncology. It is entirely possible for an otherwise well-educated oncologist to have never even heard of them.

For a variety of reasons—historical, professional and economic—Coley's toxins seem doomed to exist, for the foreseeable future, in this "twilight zone" of "questionable" cancer treatments. Every decade or so there has been a clinical trial, but no one of these, nor all of them taken collectively, have been enough to gain approval from the FDA. Is this mere happenstance (i.e., bad luck) or is there some underlying reason that could account for the failure of this treatment to gain broader acceptance?

Economic Considerations

Coley's toxins, in their classic formulation, share the characteristics—and the characteristic problem—of most natural treatments. They are unpatentable.

Of course, considerable revenue could theoretically be made from their manufacture and administration. The two bacteria involved in its manufacture are readily and freely available and the fundamental production costs of Coley's toxins are consequently low. But none of this rises to the profits that can be made from a strongly patented pharmaceutical agent.

It is a paradoxical aspect of drug development that the low cost of a new treatment can work against it in the marketplace. The reason for this is that the cost of developing a new treatment, including the expense of shepherding it through the elaborate FDA regulatory process, has become astronomical. The cost is estimated in the hundreds of millions of dollars. Pharmaceutical companies use this expense to justify charging very high amounts for new treatments, particularly anticancer agents. Prototypical is Genentech Inc.'s antiangiogenic drug, Avastin (bevacizumab), which has been approved by the FDA for the treatment of lung, colon and breast cancer. The price per patient is over $100,000 per patient per year. Based on this, Genentech had sales of Avastin in 2008 of over $2.9 billion.[267,268]

How would biological drugs like Coley's toxins fare in this situation? The intrinsic cost of producing Coley's toxins is low. The late Wayne Martin of Fairhope, Alabama, who oversaw preparation of a version of Coley's toxins that was used by some "underground" CAM practitioners, figured that the raw materials for a six months' supply came to about one dollar per patient. The starting broth, based on an inexpensive synthetic mixture, took about 30 minutes to assemble. This was then put in an incubator, where it was inoculated with the streptococcal organism. The laboratory technician then left it and came back in ten days.

The second inoculation with *Serratia marcescens* was then performed and the technician went away again for another ten days. The total labor time totalled five hours. Yet this work produces about a quart (1,000 cubic centimeters) of Coley's toxins, at a total cost of approximately US $100. Thus, the cost was around 10¢ per cc. You will remember that according to the *The American Medical Association's New and Nonofficial Drugs*, a standard dose of the toxins was 1 to 8 minims, or 0.05 to 0.5 cc. If we take the higher figure, the cost of the materials for a particular treatment with Coley's toxins comes to 5¢. Given at the highest dose, twice per week for six months (the outside limit of the treatment), the total material costs would be $2.60.

To pass muster with the FDA under 21st century conditions one must factor in the cost of producing a pharmaceutical grade product under Good Manufacturer Practices (GMP). However, such facilities exist all over the world. Unbeknownst to most Americans, many of their pharmaceuticals are now made abroad. According to the *New York Times*, "Of the 1,154 pharmaceutical plants mentioned in generic drug applications to the Food and Drug Administration in 2007, only 13 percent were in the United States. Forty-three percent were in China, and 39 percent were in India."[269] It is in the nature of generic (i.e., unpatentable) treatments such as Coley's toxins that—like antibiotics—they can be produced almost anywhere for a relatively low cost.

Of course, this $2.60 figure does not account for the cost of administering the treatment, which optimally should involve both medical and nursing supervision to prevent untoward reactions to the toxins. Nor do these calculations take into account the potentially high costs of performing further clinical trials, including randomized controlled trials (RCTs), which would almost certainly be demanded before the toxins would receive final FDA approval. This could run into millions of dollars, and it is not clear whose economic interests would be served by conducting such trials. Yet, with all that, Coley's toxins are still a potentially inexpensive agent, which almost any pharmaceutical laboratory could produce in large amounts.

It is a central fact of the present system of drug development that—ironically—this is what makes the toxins *unattractive* to those who are concerned with the bottom line of pharmaceutical companies.[270] How does one make a large profit from a drug that can be easily manufactured at virtually any laboratory in the world? The person who invests the millions to perform the clinical trials has no guarantee that he or she will reap the rewards of their efforts. Without the security of strong patents, the risk is too great, and the rewards too small to attract the major players that dominate the drug development market.

Is there a solution to this problem? Not without some modification of the present-day system of drug development, in which privately owned patents form the basis for a legal 20-year monopoly in the sale of the treatment in question. This patent-profit loop, however familiar it has become, may not an inevitable feature of drug development in the future. Some economists and legislators have recognized the scope of this problem and proposed possible solutions.

In 2006, the Nobel laureate economist Joseph Stiglitz, PhD, wrote an article in *New Scientist* in which he asserted: "It is right to reward innovation, but the way we do it now exacts too high a price from the world's poor."[271]

There are possible alternatives to the current system, however unlikely their adoptation may seem at the present time. For example, Sen. Bernie Sanders (I-VT) has proposed a bill in Congress that "would eliminate drug monopolies; instead, all medical developments would automatically enter the public domain, so more companies could manufacture each new drug. With increased manufacturing competition and zero licensing fees, drug prices should, theoretically, plummet."[272]

We would need a political and philosophical shift away from the current reliance on the patent system, and on those government-industrial partnerships that dominated the development of new drugs. Short of a radical restructuring of intellectual property rights, it is unlikely that the free market system will ever endorse Coley's toxins, although it may of course allow for the development of patented pharmaceutical agents based on Coley's insights. There is also the possibility that a wealthy private individual or foundation or a governmental agency (such as the National Center for Complementary and Alternative Medicine) might itself underwrite the developmental costs for such a natural product.

But that could not happen in a vacuum. In order for such a treatment to be a success, it would require a change in attitude among an influential section of oncologists, that such empirically validated immunomodulatory methods as Coley's toxins are worth exploring. In other words, one cannot do an 'end run' around science. A substantial number of scientists would have to *believe* that a large and rigorous clinical trial would be a productive use of research funds. We are far from such a consensus today, and one could fit the number of scientists who have even heard of Coley in a small room. The zero interest at ASCO and AACR is itself eloquent testimony to that fact. On the other hand, if Coley's toxins do sometimes work as they seem to do, then we know that they cannot be suppressed forever.

The history of Coley's toxins is alternately frustrating and inspiring. It shows that in the past there was an experimental cancer treatment available that exceeded in efficacy some of the then-current standard treatments. Some forms of advanced cancer that even today are considered incurable were sometimes put into long-term, lasting remissions via the toxins.

While the quality of the Coley trials varies greatly—and none could be considered definitive—all of them, in their own ways, showed that the treatment was active and that some Coley patients benefited from treatment, sometimes with long-lasting complete remissions. Most concluded that the treatment was promising and that further investigation was justified. Yet in almost every case there were no successful follow-up attempts to extend these findings to definitive trials.[273,274]

When we consider that many current drug trials show an actual survival benefit measured in weeks, we should be willing to vigorously investigate any treatment, no matter how unorthodox, that shows equal or greater promise. Not to do so is to betray the trust of the donors and taxpayers who support almost all cancer research.

Other Barriers to Acceptance

The history of Coley's toxins presents unusual features to students of medical history. Oftentimes, the proponents of unconventional treatments are interlopers. Some are not cancer specialists or even medical doctors. Their treatments may be promoted in various unconventional ways, such as through advertisements or, more recently, over the Internet. What makes Coley's treatment stand out is that its inventor was a highly regarded oncologist at one of the world's foremost cancer centers, with scores of scientific publications to his credit. In 40 years, he never attempted to bypass his medical colleagues and directly address the public with either his claims or his frustrations. Even his most determined professional opponent, James Ewing, reluctantly agreed that sometimes he had notable success.

Far from exploiting patients for financial gain, Coley was exceptionally generous. In 1897, he established the Needy Patient Fund at Memorial Hospital. In the following year, he treated some 40 of these charity patients from around the US and often paid their bills out of his own pocket. Coley himself performed fully 60 percent of the free surgical operations at Memorial Hospital between the years 1908 and 1913.[275] Like Roentgen with his x rays, Coley did not attempt to patent or otherwise monopolize his treatment, and in fact spent a great deal of his own money sharing his knowledge with scientists around the world. There were no secrets to Coley's methods, other than those that Nature herself kept from him and the rest of the scientific community.

Perhaps the most touching tribute came in an obituary written by Carl Burdick, MD, a prominent New York surgeon at the time:

> "We who were associated with Dr. Coley, and who recognized the sterling qualities of his rare personality, realized that there was one quality which surpassed all others, even his boundless optimism, and that was his ever-readiness—in fact, eagerness—to lend a helping hand. None appealed to him in vain. He loved his fellow men."[276]

Complementary and alternative (CAM) cancer practitioners are also often accused of disregarding the long-term outcomes of their cases. But Dr. Coley not only wrote many articles but kept copious records, which made it possible for his intrepid daughter to eventually compile the outcome of more than 1,000 cases which had been treated by either Coley himself, his son, Bradley, or their worldwide collaborators.

CAM cancer treatments are also often characterized as inordinately expensive. Sometimes that is the case (although this problem is hardly unique to alternative methods). But one of the intriguing things about Coley's toxins is that the cost of manufacturing them is potentially low. They are relatively simple to produce and could be ideally suited for use in developing countries, which today are being priced out of the 'targeted' chemotherapy market.

Helen Coley Nauts died on Jan. 2, 2001 and Coley's toxins lost its most articulate advocate, and its living link to the classic era of mixed bacterial vaccines. Because of the FDA's ban on the marketing of Coley's toxins in the US, it became a kind of "underground" treatment. One physician in Milwaukee, Wisconsin, Burton Waisbren, Sr., MD, used a mild form of the treatment as palliation for many years.[277] But he has now discontinued such use.[278] A medical writer, Wayne Martin (1912-2006), who learned about Coley's work in the 1930s, for years propagandized on their behalf in the pages of the *Townsend Letter for Doctors.* With Martin's help, an integrative physician in Florida, the late Donald Carrow, MD (1934-2003), once provided a version of Coley's toxins to cancer patients for several years.[279] Coley's toxins are also being used at a small clinic in Guatemala. But none of this moved Coley's toxins one step closer to approval.

MBVax

Today, the commercial leader in producing what it calls "Coley's fluid" is MBVax Biosciences of Ancaster, Ontario. Their product is currently used in a number of countries (other than the country of their birth). According to Don Macadam, chief executive officer of MBVax, clinics and hospitals in Austria, Bahamas, China, Germany, Ireland, Mexico, South Africa, Switzerland and Uzbekistan are using their product. In certain countries, they state, permission to treat individual cancer patients has been granted under compassionate use or on a "named-patient" basis.[280]

Donald H. MacAdam is a director of Kyto BioPharma Inc, a developer of antibody-based cancer drugs (www.kytobiopharma.com). He is a former director of CTL Immuno Therapies Corporation, a producer of therapeutic cancer vaccines. He is also the former president of CRS Robotics Corporation, a manufacturer of robotic systems for the Human

Genome Project, and Tm Bioscience Corporation, a manufacturer of biochips and developer of DNA technologies..

In 2003, disappointed with the clinical trial results of another cancer drug, MacAdam wrote a study, *Spontaneous Regression—Cancer and the Immune System,* which included a chapter on William B. Coley. That book came to the attention of radio talk show host Deborah Ray. She is the widow of Donald Carrow, MD, a physician who made available a form of Coley's toxins to his patients. After this radio appearance, Macadam received a great many letters and emails from both patients and researchers looking for a source of Coley's toxin. He sensed an opportunity.

In 2004, after further researching Coley's toxins, MacAdam thought that he had figured out how to exactly reproduce the best historical version, which, he claims, was Type XI, manufactured by Martha Tracy for the use of Dr. Coley in his own practice. MacAdam also claims to have figured out how to eliminate the batch-to-batch variations that were a drawback of all previous Coley's toxins preparations.

At this point, MacAdam came across an article, *Dr William Coley and Tumor Regression: A Place in History or in the Future*, by Prof. Stephen Hoption Cann, a medical researcher at the University of British Columbia. (It is now available for downloading free of charge at www.mbvax.com)

Hoption Cann, MacAdam, and an interdisciplinary group of 30 scientists and physicians, founded MBVax Bioscience to produce and disseminate their 'optimal' version of Coley's toxins. By the end of 2006, MBVax Coley Fluid was in production and patients were being treated at various international clinics that were authorized by their various national authoriies to employ experimental cancer therapies. By the end of 2007, patients had been treated in more than a dozen countries. In February 2008, the *Los Angeles Times* concurred: "MacAdam has solved Coley's major difficulty: maintaining consistency of the brew from batch to batch."[281]

In the meanwhile, Dr. Hoption Cann had been working with physicians and patients to optimize the clinical administration protocols and diagnostics. By the end of 2008, more than 50 advanced cancer patients had received at least 3 weeks of the new Coley Fluid therapy. According to Mr. Macadam, almost all patients have had subjective improvement with either reduced cancer pain, increased appetite, less depression and/or increased mobility. Again, according to the company, about two-thirds of patients had apparent tumor regression ("apparent" because not all patients have had access to PET/CT and other gold-standard diagnostics). Most remarkably, in five cases (three confirmed by PET/CT scans) total remission of disease was achieved (with no further evidence of cancer).

Coley Pharmaceutical Group

In addition, there are a number of companies that are trying to devise products that are not Coley's toxins, but are based on Coley's original discoveries. Most prominent of is Coley Pharmaceutical Group, with offices in Wellesley, Mass., Ottawa, Canada, and Düsseldorf, Germany. They have proprietary TLR Therapeutics™, drugs that target a subtype of the immune system called toll-like receptors (TLRs). These patented products[282], according to their Web site, "tap the natural power of the immune system, directing targeted and specific immune responses to diseases."[283] In 2007, Pfizer, Inc., which had previously cooperated with Coley in developing their "Promune" vaccine, purchased the company for US $164 million. This was a testimony to the power of Coley's ideas on cancer and immunity, more than a century after he first developed his toxin therapy.

Aldara

Actually, there is already one drug on the market that works through a toll-like receptor (similar therefore to Coley's toxins). It was developed, however, before scientists understood the role that this receptor actually played. This drug is Aldara, manufactured by 3M. It is used to treat genital herpes, as well as actinic keratosis (a precancerous skin condition) and superficial basal cell carcinoma (www.aldara.com). US Sales are $211 million per year, according to the IMS, a drug data and consulting firm.[284]

Germany and Coley's Toxins

German doctors in particular have had a long association with Coley's toxins. Most of the original observations on erysipelas and cancer were made in Germany. For a period of 67 years, from 1914 to 1981, a Munich-based company, Südmedica, produced Coley's toxins under the trade name Vaccineurin.[285] In December 1981 they ceased production of this product because of a lack of re-approval by the German Federal Institute for Drugs and Medical Devices. The celebrated German CAM practitioner Josef Issels[286] treated cancer patients with Vaccineurin for many years. In the 1980s, E. Dieter Hager, MD, PhD, founding director of the Biomed Klinik, Bad Bergzabern, and Ulrich Abel, PhD, a biostatistician in Heidelberg/Mannheim, encouraged clinical research as well as the therapeutic use of endogenous, active hyperthermia ("fever therapy") with Coley-like products.[287,xviii]

Coley's toxins use never died out in Germany and in fact it is still available there. A few clinics used their dwindling stocks of Vaccineurin, until these were depleted. Arno Thaller, MD, who uses fever therapy at his clinic in Markt Berolzheim, now uses Coley's toxins produced by OrganoMed, a Hamburg company headed by Dr. Günter Neumeyer (www.organomed.org, 011-49-4165-8981). One advantage, he says, is that the

---

[xviii] They also researched the effectiveness of the related passive "Fever-Range Whole Body Hyperthermia" (FR-WBH) using infrared radiation, a field that in part grew out of consideration of Coley's toxins.

Organomed toxins do not contain a preservative. He attributes preservatives in MBVax (i.e., thymol) with causing an allergic reactions (urticaria) in one of his patients.[288]

Conclusions

Well over 100 years have elapsed since Dr. William Bradley Coley first injected a patient with a mixed bacterial vaccine and caused a remission of her cancer. In the intervening decades, tens of millions of people have died of cancer, yet the vast majority of them were never told about Coley's toxins or other promising immunological treatments. This is tragic.

We do not have large, well-conducted randomized controlled trials (RCTs) to prove the value of Coley's toxins. But there are smaller trials, as well as many well documented case histories. The preponderance of this evidence favors the usefulness of Coley's toxins *in some instances*. It is therefore imperative that further investigations be done to delineate those situations in which Coley's vaccine might be effective. Had these tests been performed when it was first proposed in the 1920s, the history of cancer treatment might have been quite different. Protocols could have been created for the treatment of bone as well as other kinds of cancer in which this treatment seemed to offer a good chance of working.

Some researchers who are sympathetic to the immunological approach consider Coley's toxins, while advanced for their time, essentially passé. They dislike the use of crude substances without a clear scientifically defined mechanism of action. They look forward to some new era in immunotherapy, in which they themselves can give full scope to their own inventiveness and enterprise. "There is no magic to Coley's toxins," they say. John M. Timmerman, MD, an immunologist at the University of California at Los Angeles (UCLA) has written that Coley's work gave rise to "the most exciting sets of compounds in cancer immunology."[289] Yes, the development of new agents based on Coley's treatment philosophy—'new wine in old bottles'—is certainly necessary.

The trouble with this argument is that the true potential of Coley's toxins remains unknown. We do not know, for instance, what would be the result if doctors were free to treat less-than-moribund patients with an optimal version of the toxins, creating adequate weekly or biweekly fevers for four to six months. These results might surprise us. This should be the baseline for all subsequent attempts to improve on the original formula. Proposed new treatments could then be measured against this baseline, to see if they are truly superior in any regard.

Coley was an almost ideal figure for launching a new cancer treatment. He was brilliant, well educated and experienced. What is more, in the voluminous documentary record, I cannot find a single person who ever questioned his personal probity. Since he had every advantage, we can say that he was doomed to fail, given the conditions of his times. The toxins were too radical an idea and the conservative opposition too firmly entrenched. Coley was not only proposing a promising treatment, but he was conceptualizing an entire approach to cancer, which we today call immunotherapy. The appearance of other competitive treatment ideas (primarily radiation and chemotherapy) squeezed this early form of immunotherapy out of the race. It is still trying to get back in…a fourth modality that never quite gets acceptance in clinical practice.

And what about today's cancer patients? The Food and Drug Administration (FDA) erred in 1963 when it refused to "grandfather" Coley's toxins as an old and established drug, the way it did aspirin. This, however, could be rectified by giving Coley's toxins "orphan drug status," first for bone and soft tissue sarcomas, then for other cancers. This would be an acknowledgment that government intervention is required in order to motivate the free market to develop this non-patentable drug. If Coley's toxins prove effective under modern conditions for these rare cancers, they could then be approved for the more common cancers as well.

Needless to say, a new generation of doctors needs to learn about the successes (as well as the failures) of the past, sum up the lessons, and be free to utilize this promising treatment in patients who have no other legitimate choices. This would mean some sort of legislation to allow 'terminal' cancer patients to exercise their medical freedom of choice, a long-term goal for the patients rights movement. The immediate task at hand is for doctors to put Coley's toxins into clinical practice once again, this time using the advanced research tools and rigorous investigative methods of 21st century medicine. This will help not only patients of the future but will right one of the greatest wrongs in American medicine, the suppression and neglect of Coley's toxins.

Appendix A

Members of the Coley Family Mentioned in This Report

Name	Relationship	Birthplace	Born	Died
Samuel Coley (a.k.a. Coole or Cooley)	Paternal ancestor	Great Munden, Hertfordshire, England	1615 (arrived in Mass., May 1631 age 16)	17 Aug 1684
Peter Coley	Son of Samuel Coley (above)	Milford, Fairfield, CT	25 Apr 1641[xix]	31 March 1690
John Wakeman	Maternal ancestor	Bewdley, England	29 Mar 1607[*] (arrived in America, 1664)	1661
Samuel	Son of John	Fairfield, CT	07 June 1635	08 March 1692

[xix] Baptism date.

Wakeman	Wakeman			
William Coley	Paternal grandfather	Saugatuck, Westport, CT	30 Apr 1799	13 Apr 1886
Eunice (Fanton) Coley	Paternal grandmother	Saugatuck, Westport, CT	19 Nov 1798	19 Sept 1851
Horace Bradley Coley	Father	Saugatuck, Westport, CT	29 June 1829	28 Feb 1913
Clarina Bradley (Wakeman)	Mother	Fairfield, CT	15 Dec 1837	1863[290]
Abigail/Abbey Augusta (Gray) Coley	1st stepmother	Fairfield, CT	30 March 1830	27 Nov 1879
Eleanor W. (Gray) Coley	2nd stepmother	Fairfield, CT	29 Dec 1833 (Married 12 Dec. 1880)	Died between 1910 and 1920 censuses
William Bradley Coley	Self	Saugatuck, Westport, CT	12 Jan 1862	16 Apr 1936
Carrie E. Coley Gorham	Sister	Saugatuck, Westport, CT	5 Dec 1859	22 Feb 1892
Alice (Lancaster) Coley	Wife	Newton, MA	15 July 1866	29 Nov 1936
Charles Bartlett Lancaster	Father-in-law	Boston, MA	7 Jan 1837	22 Dec 1913
Mary Elizabeth (Brackett) Lancaster	Mother-in-law	Newton, MA	13 Apr 1837	2 Dec 1902
Bradley Lancaster Coley, MD	Son	New York, NY	23 Dec 1892	1 June 1961
Malcolm Coley	Son	New York, NY	29 Nov 1896	23 Sept 1901
Helen Coley Nauts	Daughter	Sharon, CT	2 Sept 1907	2 Jan 2001
William Boone Nauts, Jr.	Helen's husband	Tennessee	Oct. 23 1902	5 Apr 1990
Phyllis MacD. Coley	Bradley's wife	Vancouver, B.C., Canada	30 Aug 1898	16 June 1981

Data: From www.ancestry.com (including *Genealogical and Family History of the State of Connecticut, Vol. I-IV*, [291]), www.lybbertfamily.org,) *New York Times*, etc.

Appendix II

Key Articles of William B. Coley on his Mixed Bacterial Vaccine

Coley WB. Treatment of inoperable malignant tumors with toxins of erysipelas and the bacillus prodigiosus. *Trans Am Surg Assoc* 1894a;12:183-212.

Coley WB. The treatment of inoperable malignant tumors with the toxins of erysipelas and the bacillus prodigiosus. *Am J Med Sci* 1894b. 108: 50-66.

Coley WB. The therapeutic value of the mixed toxins of the streptococcus of erysipelas and bacillus prodigiosus in the treatment of inoperable malignant tumors. With a report of 160 cases. Am J Med Sci 1896a. 112: 251-281.

Coley WB. Further observations upon the treatment of malignant tumors with the toxins of erysipelas and Bacillus prodigiosus with a report of 160 cases. *Bull John Hopkins Hospital*. 1896b. 7: 175.

Coley WB. The treatment of malignant tumors by repeated inoculations of erysipelas: With a report of ten original cases. Am J Med Sci 1899. 105: 487-511.

Coley WB. Inoperable sarcoma: A further report of cases successfully treated with the mixed bacterial toxins of erysipelas and bacillus prodigiosus. *Med Rec*. 1907. 72: 129-137.

Coley, WB. A report of recent cases of inoperable sarcomas successfully treated with mixed toxines of erysipelas and bacillus prodigiosus. *Surg Gynec Obstetr*. 1911. 13: 174-190.

WB. Coley and BL. Coley. Primary malignant tumors of the long bones. *Archiv Surg* 1926. 13: 780-836/14. 63-141.

WB. Coley. Endothelioma myeloma of tibia: Long-standing cure by toxin treatment. *Ann Surg*. 1931. 94: 447-452.

References

[1] Pearce Gould, A. The Bradshaw lecture on cancer. *Lancet*. 1910;2:1665-1673.

---

[2] Anonymous, Vanishing cancer. *Time.* Sept. 22, 1958. Accessed April 14, 2009 from: www.time.com

[3] Wangensteen OH. Should patients be told they have cancer? *Surgery. 1950;27:944.*

[4] Lewison, Edward F. *Spontaneous Regression of Cancer.* Conference proceedings. Baltimore: Johns Hopkins Medical Institute, May 9-10, 1974.

[5] Cole WH, Everson TC. Spontaneous regression of cancer: preliminary report. *Ann Surg.* 1956;144:366-383.

[6] Conant, William Cowper. *Salvation.* 1899;1:93.

[7] Hoption Cann SA, van Netten JP, van Netten C, Glover DW. Spontaneous regression: a hidden treasure buried in time. *Med. Hypotheses.* 2002;58:115–119.

[8] Ridpath, John Clark, Ed.. *The Standard American Encyclopedia of Arts, Sciences, History, Biography.* New York: The Encyclopedia Publishing Co., 1897, p. 479.

[9] Anonymous. *Suppuration.* In: Watson, Chalmers, Ed. *Encyclopedia Medica*, New York: Longmans, Green & Co., 1901, vol. 11.

[10] O'Regan, Brendan and Hirschberg, Carolyn. *Op. cit.*

[11] Rohdenburg G. Fluctuations in the growth energy of malignant tumors in man, with especial reference to spontaneous recession. *J Cancer Res*. 1918;3:193–225.

[12] Smithers, DW. *On the Nature of Neoplasia in Man*. Edinburgh: Livingstone, 1964.

[13] Booth G. *The Cancer Epidemic.* Lewiston, NY: Edwin Mellin Press, 1979.

[14] Booth G. Booth, G. Psychobiological aspects of "spontaneous" regressions of cancer. *J. Amer. Acad. Psychoanal*. 1973;1:303-317.

[15] Everson, Tilden C and Cole, Warren H. *Spontaneous Regression of Cancer. A Study and Abstract of Reports in the World Medical Literature and of Personal Communications Concerning Spontaneous Regression of Malignant Disease.* Philadelphia and London: W.B. Saunders and Co., 1966.

[16] Cole, Warren H. 1960 American Cancer Society Presidential Address. *CA Cancer J Clin*. 1961;11:2-7.

[17] O'Regan, Brendan and Hirschberg, Caryle. *Op. cit.*

[18] Hirschberg, Caryle and Barasch, Marc Ian. *Remarkable Recovery: What Extraordinary Healings Can Tell Us about Getting Well and Staying Well.* New York: Putnam Publishing Group, 1995.

[19] Thomas, Lewis. *The Youngest Science: Notes of a Medicine-watcher*. New York: Viking Press, 1983, p. 205.

[20] Zahl P-H, Maehlen J, Welch, HG. The Natural history of invasive breast cancers detected by screening mammography. *Arch Intern Med.* 2008;168:2311-2316.

[21] *Ibid.*

[22] Mulcahy N. Mammography study suggests some breast cancer may spontaneously regress. *Medscape Medical News.* Nov 25 2008, Accessed February 16, 2009 from: http:// www.medscape.com/viewarticle/584147.

[23] Celsus II.8.17-18, cited in Coulter, Harris. *Divided Legacy*, vol. I. Washington, DC: Wehawken Book Co., 1975, vol. I, p. 256.

[24] Peck, Harry Thurston. *Harpers Dictionary of Classical Antiquities,* New York, 1898

[25] Fenaglio CG. Obstinate veneral ulcer, tending to cancer, cured by the occurrence of inflammation. *The Quarterly Journal of Foreign and British Medicine and Surgery.* 1823;5:150.

[26] Bier, August. *Hyperemia as a Therapeutic Agent*. Translated by Gustavus M. Blech. Chicago: A. Robertson & Co, 1905, p. 13. The quote is attributed to S.W. Sachs, otherwise unknown.
[27] Palmer, James F. *The Works of John Hunter*, vol. 3. London: Longman, Rees, et al., 1837, p. 418.
[28] Adams, Francis (trans.) *Genuine Works of Hippocrates.* Krieger Publishing Co., 1972. Accessed February 15, 2009 from: http://classics.mit.edu/Hippocrates/epidemics.2.ii.html
[29] Laurie, Joseph. *Homœopathic Domestic Medicine*. New York: William Radde, 1846.
[30] Hall, Stephen. *A Commotion in the Blood*, New York: Henry Holt, 1997.
[31] Zuelzer, Wilhelm. Erysipelas. In: *Cyclopedia of the Practice of Medicine.* Albert H. Buck, MD, editor of the English translation. H. von Ziemssen, ed. Vol II, Acute Infectious Diseases. London: Samson, Low, 1875.
[32] Observations on the nature and treatment of erysipelas, illustrated by cases. *Med Chir Trans.* 1828;14(Pt 1):1–80.
[33] Sabatier J. C. *Propositions Sur L'érysipèle, Considéré Principalement Comme Moyen Curatif Dans Les Affections Cutanées Chroniques: Précédées De Quelques Généralités Sur La Pathologie Cutanée Et L'influence Souvent Avantageuse Des Maladies Intercurrentes.* Paris: Didot le Jeune, 1831. [According to worldcat.org, the sole library copy is in the rare book room of Harvard Medical School.]
[34] Sabatier JC. *Recherches Historiques sur la Faculté de Médecine de Paris, Depuis son Origine jusqu' à nos Jours.* Paris, Baillière, 1837.
[35] Mauriac, Charles. *Ueber den heilenden Einfuss des Erysipels bei Syphilis*. Paris, 1874.
[36] Fehleisen F. On erysipelas. In: *Recent Essays by Various Authors on Bacteria in Relation to Disease*, Cheyne, W. Watson, ed. London: New Sydenham Society, 1886, pp. 261-288. [A readable account of Fehleisen's discovery of *Streptococcus erysipelatis*.]
[37] Després, Armand. *Traité de l'érysipèle*. Paris, Delahaye, 1862. [Copy at the NLM]
[38] Busch W. Ueber den Einfluss, welchen heftige Erysipele zuweilen auf organisirte Neubildungen ausiiben. *Berliner klin. Wochenschr.* 1866;13:245-246.
[39] Busch W. Niederrheinische Gesellschaft fur Natur-und Heilkunde in Bonn. Aus der sitzung der medizinischen section vom 13. November 1867. *Berliner Klinische Wochenschrift* 1868;5:137.
[40] Hobohm U. Healing heat: Harnessing infection to fight cancer. *Amer Sci.* 2009;97(1):34.
[41] Uwe Hobohm, personal communication, January 11, 2009.
[42] Moore JW, Immermann H, von Juergensen T, et al. Variola, Vaccination, Varicella, Cholera, Erysipelas, Whooping Cough, Hay Fever. Alfred Stengel, trans. Philadelphia: Saunders, 1902. Accessed on Feb. 9, 2009 from: http://www.archive.org/stream/variolavaccinati00mooruoft/variolavaccinati00mooruoft_djvu.txt
[43] Hall, Stephen. *op.cit.*
[44] Hobohm U. Fever therapy revisited. *British J Cancer.* 2005;92:421–425.
[45] Busch: Über den Einfluß, welchen heftige Erysipele zuweilen auf organisierte Neubildungen ausüben. *Verhandlungen des naturhistorischen Vereins der preußischen Rheinlande und Westfalens.* 1866;23:28-30.

[46] Busch W. Niederrheinische Gesellschaft fur Natur-und Heilkunde in Bonn. Aus der sitzung der medizinischen section vom 13. November 1867. *Berliner Klinische Wochenschrift* 1868;5:137.
[47] Busch W: Verhandlungen ärztlicher Gesellschaften. *Berl Klin Wochenschr.* 1866;3 :245–246.
[48] Fehleisen V. Über die Züchtung der Erysipelkokken auf künstlichem Nährboden und Ihre Übertragbarkeit auf den Menschen. *Dtsch Med Wochenschr.* 1882.8:533-554.
[49] Sternberg, George M. *A Textbook of Bacteriology*. 2nd ed. New York: William wood and Co., 1901, p. 382.
[50] Rosenbach AFJ. Über das Erysipeloid. *Verhandlungen der Deutschen Gesellschaft für Chirurgie,* 1887.
[51] Fehleisen V. *On Erysipelas*. Translated by Leslie Ogilve. In: W.W. Cheyne, editor: *Microparasites in Disease, Selected Essays*. London, 1886, p. 272. Cited in Bainbridge, William Seaman. *The Cancer Problem*. New York: Macmillan, p. 333.
[52] Hobohm U. Healing heat: Harnessing infection to fight cancer. *Amer Sci.* 2009;97(1):34.
[53] Fehleisen V. VI. Über die Züchtung der Erysipelkokken auf künstlichem Nährboden und Ihre Übertragbarkeit auf den Menschen. *Dtsch Med Wochenschr* 1882. 8: 533-554.
[54] Bruns P. Die Heilwurkung des Erysipels auf Geschwülste. *Beitrage f Klin.Chir.* 1888;3:443-466.
[55] Richter PF. Leukemia and Erysipel. *Charite-annalens*. 1896. 21: 299-309.
[56] Petruschky J. Entscheidungsversuche zur Frage der Specifitat des Erysipel-streptococcus. *Zeitschr. f. Hygiene und Infectionskrankheiten*. Bd. xxm.
[57] Petruschky J. Die verschiedenen Erscheinungsformen der Streptococceninf ection," etc., ebenda, Bd. xvin.
[58] Volkmann, Richard von. Behandlung des Erysipelas, des Lupus, der Synovitis.
[59] Isham AB. The influence of typhoid fever upon venereal diseases. *Cincinnati Lancet-Clinic.* 1890; n.s. 25:123-125.
[60] R. Eschweiler. *Die Erysipeltoxin und Serumtherapie der bösartigen Geschwülste.* Leipzig: C.G. Naumann, 1897. [I haven't seen this very rare work and am relying on a summary given by Bradley L. Coley, MD]
[61] Gresser IA. Chekhov, M.D., and Coley's toxins. *N. Engl. J. Med.* 1987;317:457.
[62] Chekhov, Anton Pavlovich. *Letters of Chekhov to his Family and Friends*. Constant Garnett, trans. New York: The Macmillan Co., 1920, p. 221.
[63] Wagner-Jauregg, Julius von. Fieber- und Infektionstherapie; ausgewählte Beiträge, 1887-1935. Wien: Weidmann, 1936.
[64] Accessed April 14, 2009 from: http://newswire.rockefeller.edu/index.php?page=engine&id=621
[65] Accessed April 14, 2009 from: http://familytreemaker.genealogy.com/users/p/o/p/Sandra-Popiel/FILE/0750text.txt
[66] Wakeman, Robert Peel. *Wakeman Genealogy, 1630-1899*. Meriden, Connecticut: Journal Pub. 1900, p. 75.
[67] Hall, Stephen. *Op.cit.*
[68] Walsh, James J. *History of Medicine in New York: Three Centuries of Medical Progress*. New York: National Americana Society, 1919, p. 241.

---

[69] Sweetser, M.F. *King's Handbook of Newton*. Boston: Moses King Corp., 1889, p. 110.
[70] Levine DB. The Hospital for the Ruptured and Crippled: William Bradley Coley, Third Surgeon-in-Chief 1925-1933. *HSS J.* 2008;4:1-9.
[71] *New York and New England Association of Railway Surgeons. Transactions of the 31st Annual Session*. Binghamton, NY: Chaffee, George, 1921.
[72] Dennis, Frederck S. *System of Surgery*. Philadelphia: Lea Bros, 1895.
[73] Warren, John Collins; Gould, Alfred Pearce, Sir. The International Text-book of Surgery. Philadelphia: Saunders, 2nd ed. 1902.
[74] Keen, William W. *Surgery: Its Principles and Practice*. Philadelphia: Saunders, 1906-1921.
[75] Coley WB. Article on abdominal surgery. In: Hare MD and Landis HRM, eds. *Progressive Medicine*, vol. 2. Philadelphia: Lea Bros., 1904.
[76] Anonymous. Review. In: Hare HA and Martin E. *Therapeutic Gazette*. 1902;18:497-498.
[77] Bassini, Edoardo. *Nuovo Metodo Operativo per la Cura Radicale dell'ernia Inguinale*. Padova: Prosperini, 1889. [A new operative method for the radical cure of inguinal hernia.]
[78] Lister, Lord Joseph. *Br Med J* 1897 April 3; 1(1892): 861–862.
[79] Keen, William Williams. *Animal Experimentation and Medical Progress*. Boston and New York: Houghton Mifflin, 1914, p. 193-194. [Keen is quoting Lord Joseph Lister.]
[80] Cullen, Thomas Steven. *Cancer of the Uterus*, New York: D. Appleton and Co, 1900, p. 651.
[81] Coley WB. *Annals of Surgery* 1891;14:199-220.
[82] Hall, *op.cit.,* p. 41.
[83] *Ibid.*
[84] *Ibid.*
[85] Coley WB. Contribution to the knowledge of sarcoma. *Annals of Surgery.* 1891;14:199-200.
[86] Hall, op. cit., p. 42
[87] Hobohm U. Healing heat: Harnessing infection to fight cancer. *Amer Sci.* 2009;97(1):34.
[88] Nauts, Helen Coley. Letter to the author, March 22, 1991.
[89] *Ibid.*
[90] Arrowsmith H, Porter WH, Sondern FE. The death of Dr. Farquhar Ferguson. *Yearbook of the Medical Association of the Greater City of New York*. New York: Medical Association of the Greater City of New York, 1908, p. 32-33.
[91] Coley WB. Contribution to the knowledge of sarcoma. *Annals of Surgery.* 1899;14:199.
[92] Coley WB. Late results of treatment of inoperable sarcoma by the mixed toxins of Erysipelas and Bacillus prodigiosus. *American Journal of the Medical Science*, March 1906.
[93] Coley WB. Late results of the treatment of inoperable sarcoma by the mixed toxins of Erysipelas and Bacillus prodigosus. *Reports of the Collis P. Huntington Fund for Cancer Research*, New York City, March, 1906, p. 1-56.

---

[94] Accessed April 14, 2009 from:
http://findarticles.com/p/articles/mi_m0ISW/is_/ai_97994382

[95] Piccirillo JF, Costas I, Reichman, ME. Cancers of the Head and Neck. In Ries LAG, Young LF, Keel GE, Eisner MP, Lin YD, Horner M-J (eds): SEER Survival Monograph: Cancer Survival Among Adults: U.S. SEER Program, 1988-2001, Patient and Tumor Characteristics. National Cancer Institute, SEER Program, NIH Pub. No. 07-6215, Bethesda, MD, 2007, chapter 2.

[96] Hobohm U. Healing heat: Harnessing infection to fight cancer. *Amer Sci.* 2009;97:34.

[97] Roger GH. *Compt rend Soc de Biol.* Paris 1890;42:573-580.

[98] *Roger GH. Principles of Medical Pathology.* Trans. M.S. Gabriel. New York and London: D. Appleton, 1905, pp. 143-144.

[99] Coley WB. Value of the mixed toxins of erysipelas and Bacillus prodigiosus in inoperable sarcoma, based upon a study of cases treated during the past sixteen years. *Detroit Medical Journal.* 1909;9:197-205.

[100] Vaughan, Victor and Novy, Frederick G. *Cellular Toxins, or the Chemical Factors in the Causation of Disease.* 4th ed. Lea Brothers: Philadelphia and New York, 1902.

[101] Coley WB. Value of the mixed toxins of erysipelas and Bacillus prodigiosus in inoperable sarcoma, based upon a study of cases treated during the past sixteen years. *Detroit Medical Journal.* 1909;9:197-205.

[102] Hartwell JL, Shear MJ, Adams JRJ. Chemical treatment of tumours. VII. Nature of the hemorrhage-producing fraction from Serralia marcescens (Bacillus prodigiosus) culture filtrate. *J. Natl Cancer Inst.* 1943; 4:107–22.

[103] Stephen Hoption Cann, personal communication, January 29, 2009.

[104] Wang FH, Li YH, Li S, Jiang WQ, Guan ZZ, [Phase I clinical trial of intravenous recombinant human lymphotoxin-alpha derivative]. Ai Zheng. 2006 Apr;25(4):501-4 (article in Chinese).

[105] Deorukhkar AA, Chander R, Ghosh SB, Sainis KB. Identification of a red-pigmented bacterium producing a potent anti-tumor N-alkylated prodigiosin as Serratia marcescens. Res Microbiol. 2007;158:399-404.

[106] McCarthy EF. The toxins of William B. Coley and the treatment of bone and soft-tissue sarcomas". The Iowa orthopaedic journal 2006;26:154–158.

[107] Haux J. Infection and cancer. Lancet. 2001;358:155-156.

[108] Tsung K, Norton JA. Lessons from Coley's Toxin. *Surg Oncol.* 2006;15:25-28.

[109] Coley WB. The treatment of sarcoma with the mixed toxins of erysipelas and bacillus prodigiosus. *Boston Med Surg J.* 1908;158:175-182.

[110] Matagne H. Premiers essais de traitement des tumeurs malignes inoperables par les toxines de Coley: un cas de cancer gueri. *Gaz. Med. Liege.* 1896;8:401-402.

[111] Rapport de la commission gui a eté chargee d'examiner le memoire de M. le Dr. H. Matagne, a Bruxelles, intitule: Premiers essais traitement des tumeurs malignes inoperables par le toxines de Coley: un cas de cancer gueri. *Bull. Acad. roy. de Med. de Belg.* (Brussels) 1896;10:275-278. [First treatment attempts of inoperable malignant tumors by Coley's toxins: a case of cancer cured.]

[112] Matagne H. Traitement des tumeurs malignes inoperable par l'erysipele et par les toxines de Coley. Presse Med. Belge (Brussels). 1899;51: 603-607; 624-629; 633-639. [Treatment of inoperable malignant tumors by erysipelas and Coley's toxins.]

[113] Matagne H.Les toxines de Coley employees dans le but de prevenir la recidive du cancer. Presse Med. Belge, 1901;54:1-3. [Coley's toxins employed with the aim of preventing the recurrence of cancer.]
[114] Matagne, H.: Presentation de cancereux gueris par les toxines de Coley, employees conjointement avec intervention chirurgicale. Presse Med, Belge. 1905;57:173-179. [Presentatio of cancer cured by Coley's toxins, employed alongside surgical intervention.]
[115] Matagne, J.H.J.: Vers la guerison du cancer. Le Scalpel. 1953;104: 504-544. 1951; and 106; 1387-1395 [Towards the cure of cancer].
[116] Coley WB. A lifetime in American medicine. *Yale Alumni Weekly*. Jan. 20, 1933. [Contains report on Coley's retirement banquet at the Waldorf-Astoria.]
[117] Nauts, Helen Coley. Coley's toxins—The first century. *Townsend Letter for Doctors and Patients*, June 2004.
[118] Anonymous. Mrs. William B. Coley [obituary]. *New York Times*, Nov. 30, 1936.
[119] Hall, Stephen. *Op.cit*.
[120] Accessed April 14, 2009 from: www.emedicinehealth.com
[121] Herndon, Richard and Burton, Richard. *Men of Progress*. Boston: New England Magazine, 1898, p. 118.
[122] Hall, Stephen. *Op cit*.
[123] Greene, Robert. The treatment of syphilis by means of the toxins of erysipelas and bacillus prodigiosus. *The Medical News*. 1896;69:399.
[124] Reed, Harvey. Sarcoma of the upper jaw. *Columbus Medical Journal*. 1897;18:314.
[125] Coley WB. The diagnosis and treatment of bone sarcoma. *Glasgow Med J*. 1936;8:49.
[126] Boyse EA, Old LJ The immunogenetics of differentiation in the mouse. Harvey Lect. 1978;71:23-53.
[127] Anonymous. The significance of fever I. *The Boston Medical and Surgical Journal*, 1890;123:329-331.
[128] Anonymous. The significance of fever II. The utility of antipyretics. *The Boston Medical and Surgical Journal*. 1890;123:401-403.
[129] Harmer, T.W.; A study of the efficiency of mixed toxins (Coley) in inoperable sarcoma. A critical analysis of 134 microscopically proven cases. Boston M. & S. J. 172; 331-338; 373-377; 411-416; 440-448. 1915.
[130] Nauts, Helen Coley. Coley's toxins—the first century. *Townsend Letter for Doctors and Patients,* June 2004.
[131] Harmer, T.W.: Remarks upon the effects observed in the use of the mixed toxins (Coley) in certain cases of sarcoma. *Boston Med & Surg J*. 1914;171:253-261.
[132] Coley WB. Treatment of inoperable sarcoma with the mixed toxins of erysipelas and bacillus prodigiosus. *Brooklyn Med J*. 1900;14:637-645.
[133] Coley, WB. The erysipelas toxins—a reply to the editorial of Dec. 15, 1894. *J Amer Med Assoc*. 1894;24:28.
[134] *Ibid.*
[135] Bainbridge, William Seaman. *The Cancer Problem*. New York: Macmillan, 1914, p. 336.
[136] *Ibid.*

[137] Beebe SP and Tracy M. The treatment of experimental tumors with bacterial toxins. *JAMA*. 1907;49:1493-1498.
[138] Beebe SP and Tracy M. The treatment of experimental tumors with bacterial toxins. *JAMA*. 1907;49:1493-1498.
[139] Eve, Sir Frederic. Remarks on the treatment of sarcoma of the long bones. *The Lancet.* 1912 (Nov. 16);2:1355.
[140] Ibid.
[141] Bainbridge, William Seaman. *The Cancer Problem.* New York: Macmillan, 1914, p. 295
[142] Walther HWE. Tumors of the bladder. *Ann Surg*. 1917;66:682-701.
[143] Anonymous. Discussion on papers by Drs. Leonard, Boggs and Pfahler. *The Pennsylvania Medical Journal.* 1906;35:652-654.
[144] Harmer, TW. Remarks upon the effects observed in the use of the mixed toxins (Coley) in certain cases of sarcoma. *Boston M. & S. J.* 1914;171: 253-261.
[145] Nauts, Helen Coley. Coley's toxins: the first century. *Townsend Letter for Doctors and Patients,* June, 2004.
[146] McCarthy EF. The toxins of William B. Coley and the treatment of bone and soft-tissue sarcomas. *Iowa Orthop J.* 2006;26:154-158.
[147] Montague, WL. *Biographical Record of the Alumni and Non-graduates of Amherst College (Classes of 1871-1896),* vol. II. Amherst, MA: Carpenter & Morehouse, 1901, p. 309.
[148] Anonymous. James Ewing. *The Encyclopedia America: A Library of Universal Knowledge.* New York and Chicago: The Encyclopedia America Corp., 1918, p. 616.
[149] http://www.whonamedit.com/doctor.cfm/2094.html
[150] Anonymous. *New York Supplement*, vol. 137. St. Paul: West Publishing, 1913, p. 434.
[151] www.facial-neuralgia.com
[152] Accessed April 14, 2009 from: http://www.whonamedit.com/doctor.cfm/2094.html
[153] Ewing J. The cancer research hospital. Read at the opening exercises of the State Institute for the Study of Malignant Diseases, Buffalo, NY. Nov. 1, 1913. Documents of the Senate of the State of New York, 1914.
[154] Mostofi, Seyed Behrooz. *Who's Who in Orthopedics*. London: Springer-Verlag, 205, p. 72.
[155] Babbs AG. Some modern tendencies in medical education. Guy's Hospital Gazatee. 1889;3:236.
[156] Coley WB. The treatment of inoperable sarcoma by bacterial toxins (the mixed toxins of the streptococcus of erysipelas and the bacillus prodigiosus). *The Practitioner*. 13, 1909;83: 591-613.
[157] Hess, David J. *Can Bacteria Cause Cancer? Alternative Medicine Confronts Big Science.* New York: New York University Press, 1997.
[158] Coley WB. The cancer symposium at Lake Mohonk. *Amer J Surg* (n.s.). 1926;1:222-225.
[159] Ewing J. *Neoplastic Diseases*, 1st ed. Philadelphia: Saunders, 1914, p. 114.
[160] http://www.cancer.gov/cancertopics/factsheet/hpylori
[161] David Hess, PhD, personal communication, January 8, 2009.

[162] Hess, David J. *Can Bacteria Cause Cancer? Alternative Medicine Confronts Big Science.* New York: New York University Press, 1997
[163] Ewing J. The treatment of cancer on biological principles. *New York Medical Journal* Oct. 19, 1912.
[164] Hall, Stephen. *Op.cit.*, p. 37.
[165] Ewing J. *Neoplastic Diseases*, 4th ed., Phila.: Saunders, 1941, p. 314.
[166] Coley WB. A lifetime in American medicine. *Yale Alumni Weekly*. Jan. 20, 1933.
[167] Hall, p. 89.
[168] Moss, Ralph W. *The Cancer Industry*. New York: Equinox Press, 1997.
[169] http://nobelprize.org/nobel_prizes/medicine/laureates/1966/rous-bio.html
[170] Anonymous. Article on Coley. *Who's Who in New York City and State: A Biographical Dictionary of Contemporaries.* New York: W.F. Brainard, 1911, p. 195.
[171] Nauts, Helen Coley. Letter to the author, March 22, 1991.
[172] Walsh, James Joseph. *History of Medicine in New York*, p. 242.
[173] Anonymous. Erysipelas germs as cure for cancer. *New York Times*, July 28, 1908, p. 3.
[174] Anonymous. The failure of the erysipelas toxins. *JAMA*. 1894;24:919. Editorial.
[175] http://www.time.com/time/magazine/article/0,9171,847196,00.html?iid=chix-sphere
[176] Anonymous. Good old fluid. *Time,* Oct. 22, 1934. Accessed at: http://www.time.com/
[177] Anonymous. Medical leaders at Coley's funeral. New York Times, April 19, 1936.
[178] Levine DB. The Hospital for the Ruptured and Crippled: William Bradley Coley, Third Surgeon-in-Chief 1925-1933. *HSS J.* 2008;4:1-9.
[179] Coley, Bradley Lancaster, *Neoplasms of Bone and Related Conditions; their Etiology, Pathogenesis, Diagnosis, and Treatment.* New York: Hoeber, 1949.
[180] Moss, Ralph W. The Cancer Industry, New York: Equinox Press, 1997.
[181] Nauts, Helen Coley. Letter to the author, March 22, 1991.
[182] Helen Coley picks her bridal party. *New York Times*, July 25, 1928, p. 12.
[183] Marge McAvoy, personal communication, February 6, 2009.
[184] Nauts, HC, Swift WE, Coley, BL. The treatment of malignant tumors by bacterial toxins as developed by the late William B. Coley, M. D., reviewed in the light of modern research. Cancer Res. 1946;6:205-216.
[185] Nagourney, Eric. Helen C. Nauts, 93, champion of her father's cancer work. *New York Times,* January 9, 2001.
[186] Nauts HC. *Beneficial Effects of Immunotherapy (Bacterial Toxins) on Sarcoma of the Soft Tissues, Other than Lymphosarcoma*. Monograph #16. New York: Cancer Research Institute, 1975.
[187] Old LJ, Clarke DA, Benacerraf B. Effect of Bacillus Calmette-Guerin infection on transplanted tumours in the mouse. Nature. 1959 Jul 25;184(Suppl 5):291-2.
[188] Morales A, Eidinger D, Bruce AW. Intracavitary Bacillus Calmette-Guerin in the treatment of superficial bladder tumors. *J Urol.* 1976;116:180-183.
[189] Old, LJ. Tumor immunology: the first century. *Current Biology*. 1992;4:603-607.
[190] Accessed April 14, 2009 from: www.coleytoxins.com
[191] Anonymous. Unproven methods of cancer treatment: Coley's mixed toxins. *CA Cancer J Clin.* 1965 May-Jun;15:139-40.

[192] Oldham RK. Biotherapy: the fourth modality of cancer treatment. J Cell Physiol Suppl. 1986;4:91-9. Review.
[193] Löwy, Ilana. *Between Bench and Bedside.* Cambridge, MA: Harvard University Press, 1996, p. 122.
[194] Nauts, Helen C; Fowler, George A; Pelner, Louis. Sarcoma of the soft tissues, other than lymphosarcoma, treated by toxin therapy; end results in 186 determinate cases with microscopic confirmation of diagnosis: 49 operable, 137 inoperable. New York, 1959.
[195] Fowler, George A. Beneficial effects of acute bacterial infections or bacterial toxin therapy on cancer of the colon or rectum. New York: Cancer Research Institute, 1969.
[196] Fowler, George A. Enhancement of natural resistance to malignant melanoma with special reference to the beneficial effects of concurrent infections and bacterial toxin therapy. New York: Cancer Research Institute, 1969.
[197] Fowler, George A. Testicular cancer treated by bacterial toxin therapy as a means of enhancing host resistance; end results in 63 determinate cases with microscopic confirmation of diagnosis: 20 operable (85 [per cent] successes); 26 inoperable (35 [per cent] successes); 17 terminal (6 [per cent] successes). New York, 1968.
[198] Nauts, Helen C and Fowler, George A. End results in lymphosarcoma treated by toxin therapy alone or combined with surgery and/or radiation or with concurrent bacterial infection. New York: Cancer Research Institute, 1969.
[199] Fowler, George A; Nauts, Helen C. The Apparently Beneficial Effects of Concurrent Infections, Inflammation or Fever, and of Bacterial Toxin Therapy on Neuroblastoma. New York: Cancer Resarch Institute, 1970.
[200] Med. Press 1955:234:206-208.
[201] Baron, S., ed. *Medical Microbiology*, 4th ed., U. of Texas, Galveston, 1996:348.
[202] Proc Royal Soc Med, Surgical Section. 1909-1910:1-48
[203] Coley WB. The treatment of sarcoma with the mixed toxins of erysipelas and *Bacillus prodigiosus*. Boston Med Surg J 1908;158:175.
[204] Beebe SP, Tracy M. The treatment of experimental tumors with bacterial toxins. *JAMA*. 1907; 49:1493-1498.
[205] Black KL, Ciacci JD, Ammirati M, Selch MT, Becker DP. Clinical trial of Serratia marcescens extract and radiation therapy in patients with malignant astrocytoma. *J Clin Oncol.* 1993;11:1746-1750.
[206] Coley WB. Value of the mixed toxins of erysipelas and Bacillus prodigiosus in inoperable sarcoma, based upon a study of cases treated during the past sixteen years. *Detroit Medical Journal*. 1909;9:197-205.
[207] MacAdam, DH. *Bacterial Substances and Cytokines*. Accessed April 14, 2009 from the MBVax Bioscience web site (www.MBVax.com).
[208] Montaner B, Navarro S, et al. Prodigiosin from the supernatant of Serratia marcescens induces apoptosis in haematopoietic cancer cell lines. *Br J Pharm*. 2000 131, 585-593.
[209] Nauts, Helen Coley. Coley's toxins – The first century. Townsend Letter for Doctors and Patients, June 2004.
[210] *Ibid.*
[211] Coley WB. The treatment of inoperable sarcoma with bacterial toxins. Trans New Hamp Med Soc. 1910;119:225-268.

[212] Nauts HC and Coley B. Treatment of malignant tumors by Coley bacterial toxins, in *Approaches to Tumor Chemotherapy*, ed. F.R. Moulton. Washington, DC: American Association for the Advancement of Science, 1947, pp. 217-243.
[213] Stephen Hoption Cann, personal communication, January 29, 2009.
[214] Tsung K, Norton JA. Lessons from Coley's toxin. *Surg Oncol*. 2006;15:25-28.
[215] Thotathil Z, Jameson MB Early experience with novel immunomodulators for cancer treatment. *Expert Opin Investig Drugs*. 2007;16:1391-1403.
[216] Jurk M, Vollmer J. Therapeutic applications of synthetic CpG oligodeoxynucleotides as TLR9 agonists for immune modulation. *BioDrugs*. 2007;21:387-401.
[217] Verhoef C, de Wilt JH, ten Hagen TL, Eggermont AM. Isolated hepatic perfusion for the treatment of liver tumors: sunset or sunrise? Surg Oncol Clin N Am. 2008;17:877-894, xi.
[218] Baldwin, RW. In: Steffin, C and Ludwig, H., Eds. *Clinical Immunology*, Amsterdam: Elsevier/North-Holland Biomedical Press, 1981.
[219] Prehn, RT. The immune reaction as a stimulator of tumor growth. *Science*. 1972;176:170-171. Cited in Nauts, HC. *Breast Cancer*; 1984.
[220] Ralf Kleef, MD, personal communication, January 23, 2009.
[221] Kataki A, Scheid P, Piet M, et al. Tumor infiltrating lymphocytes and macrophages have a potential dual role in lung cancer by supporting both host-defense and tumor progression. *J Lab Clin Med*. 2002;140:320-328.
[222] Nauts, Helen Coley. Breast cancer : immunological factors affecting incidence, prognosis, and survival. New York: Cancer Research Institute, 1984.
[223] Merck & Co. *The Merck Manual of Diagnosis and Therapy,* Rahway, NJ: Merck, Sharpe and Dohme, 16th ed., 1992, p. 9.
[224] Heppner, GH and Fulton AM. *Macrophages and Cancer.* Boca Raton: CRC Press, 1988, p. 150.
[225] Ralf Kleef, MD, personal communication, January 7, 2009.
[226] Hall, Stephen, *Op. cit.,* p. 104.
[227] Coley WB. Toxins in operable cases of bone sarcoma. Glasgow Med J. 1936(Aug); 126(2):49-86.
[228] Hall, Stephen, *op.cit.*, p. 105.
[229] Nauts, Helen Coley. Coley's toxins: the first hundred years. *Townsend Letter for Doctors and Patients,* June 2004.
[230] Hollon, Tom. Coley toxins hidden message. *The Scientist*. 2001;15(5):19.
[231] Ferrara BE. Hemicorporectomy: A collective review. *J Surg Oncol*. 1990;45:270–278.
[232] Anonymous. Good old fluid. *Time*. Oct. 22, 1934.
[233] Miller JF, Moreman KG. The role of the thymus in immunity *Med Biol Illus*. 1963;13:146-51.
[234] Hess, David J. *Can Bacteria Cause Cancer? Alternative Medicine Confronts Big Science.* New York: New York University Press, 1997.
[235] Hollister LE, Page IH, Pfeiffer CC, Visscher MB The Kefauver-Harris amendments of 1962: a critical appraisal of the first five years. J Clin Pharmacol J New Drugs. 1968;8:69-73.
[236] Johnston B. Clinical effects of Coley's Toxins. I. Controlled study. 2. A seven-year study. *Cancer Chemother Rep*. 1962;21:19-68.

[237] Ibid.
[238] Johnston BJ, Novales ET. Clinical effect of Coley's toxin. II. A seven-year study. *Cancer Chemother Rep*. 1962;21:43-68.
[239] Chandler JJ, Stark DB, Allen CV, Fletcher WS. Treatment of cancer by bacterial toxins. *Am Surg*. 1965;31:443-449.
[240] Havas HF, Axelrod RS, Burns MM, Murasko D, Goonewardene M. Clinical results and immunologic effects of a mixed bacterial vaccine in cancer patients. Med Oncol Tumor Pharmacother. 1993;10(4):145-58
[241] Havas HF, Donnelly AJ. Mixed bacterial toxins in the treatment of tumors. IV. Response of methylcholanthrene-induced, spontaneous, and transplanted tumors in mice. Cancer Res. 1961 Jan;21:17-25.
[242] Havas HF, Donnelly AJ, Levine SI. Mixed bacterial toxins in the treatment of tumors. III. Effect of tumor removal on the toxicity and mortality rates in mice. Cancer Res. 1960 Apr;20:393-6.
[243] Donnelly AJ, Havas HF, Groesbeck ME. Mixed bacterial toxins in the treatment of tumors. II. Gross and microscopic changes produced in sarcoma 37 and in mouse tissues. Cancer Res. 1958 Feb;18(2):149-54.
[244] Havas HF, Groesbeck ME, Donnelly AJ. Mixed bacterial toxins in the treatment of tumors. I. Methods of preparation and effects on normal and sarcoma 37-bearing mice. Cancer Res. 1958 Feb;18(2):141-8.
[245] Havas HF, Axelrod RS, Burns MM, Murasko D, Goonewardene M. Clinical results and immunologic effects of a mixed bacterial vaccine in cancer patients. Med Oncol Tumor Pharmacother. 1993;10(4):145-58
[246] Ibid.

[247] Axelrod RS, Havas HF, Murasko DM, Bushnell B, Guan CF. Effect of the mixed bacterial vaccine on the immune response of patients with non-small cell lung cancer and refractory malignancies. Cancer 1988;61:2219-30.

[248] Kempin S, Cirrencione C, Myers J, et al. Combined modality therapy of advanced nodular lymphomas (NL): The role of nonspecific immunotherapy (MBV) as an important determinant of responses and survival. Proc Am Soc Clin Oncol 1983;24:56. Meeting abstract.
[249] *Ibid.*
[250] Tang ZY, Zhou HY, Zhao G, Chai LM, Zhou M, Lu JZ, et al. Preliminary result of mixed bacterial vaccine as adjuvant treatment of hepatocellular carcinoma. Med Oncol & Tumor Pharmacother 1991;8:23-28.
[251] Zhao YT, et al. Preliminary reult of mixed bacterial vaccine as adjuvant treatment of hepatocellular carcinoma. Med Oncol & Tumor Pharmacother. 1991;8:23-28.
[252] Kölmel, K.F., Vehmeyer, K., Göhring, E., Kuhn, B. and Wieding, J.U. Treatment of advanced malignant melanoma by a pyrogenic bacterial lysate. A pilot study. Onkologie 1991;14:411-417.

[253] Bajenov LG, Nasrullaev AM, Bajenova TL. Analgesic effect of the analogue to Coley's vaccine [abstract]. 2nd World Congress on Regional Anaesthesia and Pain Therapy, Rio de Janeiro, March 2006.
[254] Accessed April 13, 2009 at: http://www.mbvax.com/publications.htm
[255] http://www.clinicaltrials.gov Accessed December 10, 2008.
[256] Don MacAdam, personal communication, January 1, 2009.
[257] Wiemann B, Starnes CO. Coley's toxins, tumor necrosis factor and cancer research: a historical perspective. Pharmacol Ther. 1994;64(3):529-564.
[258] Richardson MA, Ramirez T, Russell NC, Moye LA. Coley toxins immunotherapy: a retrospective review. *Altern Ther Health Med.* 1999 May;5(3):42-7.
[259] Bickels J, Kollender Y, Merinsky O, Meller I. Coley's toxin: historical perspective. Isr Med Assoc J. 2002 Jun;4(6):471-2.
[260] Kleef R, Jonas WB, Knogler W, Stenzinger W Fever, cancer incidence and spontaneous remissions. Neuroimmunomodulation. 2001;9(2):55-64.
[261] Hobohm U. Fever and cancer in perspective. *Cancer Immunol Immunother.* 2001;50:391-396.
[262] Uwe Hobohm, personal communication, January 9, 2009.
[263] Hobohm U. Fever therapy revisited. *Brit J Cancer.* 2005;92:421-425.
[264] Hobohm U, Stanford JL, Grange JM. Pattern-associated molecular pattern in cancer immunotherapy. *Critical Reviews in Immunotherapy.* 2008;28:95-107.
[265] McCarthy EF. The toxins of William B. Coley and the treatment of bone and soft-tissue sarcomas. Iowa Orthop J. 2006;26:154-158.
[266] Accessed April 13, 2009 at: http://www.cancerdecisions.com/061707.html
[267] http://www.gene.com/gene/ir/financials/historical/avastin.html
[268] Kolata G and Pollack A. Costly cancer drug offers hope, but also a dilemma. New York Times, July 6, 2008.
[269] Harris, Gardiner. Drug making's move abroad stirs concerns. *New York Times*, Jan. 19, 2009.
[270] Accessed April 13, 2009 at: http://www.fiercepharma.com/story/breast-cancer-uptake-boosts-avastin-sales/2008-07-01
[271] Stiglitz, Joseph. Give prizes not patents. *New Scientist.* Sept. 16, 2006.
[272] Accessed April 13, 2009 at: http://www.sanders.senate.gov/news/record.cfm?id=291042
[273] Starnes CO. Coley's toxins in perspective. *Nature.* 1992;357(6373):11-12.
[274] Richardson MA, Ramirez T, Russell NC, Moye LA. Coley toxins immunotherapy: a retrospective review. *Altern Ther Health Med.* 1999;5:42-47.
[275] Nauts, HC. *Coley's Toxins: 1893-1995 and Beyond.* Speech to Pharmacia. Lund, Sweden, May 15, 1995.
[276] Burdick CG. William Bradley Coley. *Ann Surg/* 1937:105:152-55.
[277] Waisbren, BA. Observations on the combined systemic administration of mixed bacterial vaccine, Bacillus Calmette-Guerin, transfer factor, and lymphoblastoid lymphocytes to patients with cancer, 1974-1985. *J. Biol Response Modifiers.*1987;6:1-19.
[278] Burton Waisbren, personal communication, September, 2008.
[279] Accessed April 15, 2009 from: http://www.townsendletter.com/June2008/doncarrow0608.htm

[280] Accessed March 15, 2009 from: http://www.mbvax.com/
[281] Borrel, Brendan. Cancer and the bacterial connection. *Los Angles Times*, February 18, 2008.
[282] Accessed April 15, 2009 from: http://www.freepatentsonline.com/y2007/0232622.html?query=coley+TLR+cancer&stemming=on
[283] Accessed March 1, 2009 from: http://www.coleypharma.com/about-us/index.cfm
[284] Pollack, Andrew. *A revival for immunity.* New York Times, October 5, 2005
[285] Hess, David J. *Can Bacteria Cause Cancer? Alternative Medicine Confronts Big Science.* New York: New York University Press, 1997.
[286] Eurekah Bioscience Collection Oncology Fever, Pyrogens and Cancer.
[287] Hager ED, Abel U, eds. *Biomodulation und Biotherapie des Krebses II. Endogene Fiebertherapie und exogene Hyperthermie in der Onkologie*. Heidelberg: Verlag für Medizin Dr. E. Fischer, 1987
[288] Arno Thaller, personal communication, January 26, 2008.
[289] Accessed April 16, 2009 from: http://articles.latimes.com/2008/feb/18/health/he-lab18
[290] Wakeman, Robert Peel. *Wakeman Genealogy, 1630-1899*. Meriden, Connecticut: Journal Pub. 1900.
[291] Cutter, William Richard. *Genealogical and Family History of the State of Connecticut: A Record of the Achievements of Her People in the Making of a Commonwealth and the Founding of a Nation.* New York, NY: Lewis Publishing Company, 1911.

Made in United States
Orlando, FL
05 November 2026

90231087R00056